COMMUNICATIONS FOR MANUFACTURING

COMMUNICATIONS FOR MANUFACTURING

Proceedings of the Open Congress
4–7 September 1990   Stuttgart, Germany
CEC DG XIII: Telecommunications, Information Industries and Innovation

Edited by

S. Withnell and W. Van Puymbroeck

Springer-Verlag
London Berlin Heidelberg New York
Paris Tokyo Hong Kong

Open Communications Congress

An international event organised by:
CIM-Europe, Commission of the European Communities
200 Rue de la Loi, 1049 Brussels, Belgium

Published by: Springer-Verlag, Berlin Heidelberg New York London Paris Tokyo
Printed in Great Britain by Antony Rowe Ltd

Stephen Withnell
British Aerospace (Military Aircraft) Ltd.
BAeCAM, Guild Centre Offices
Lords Walk
Preston,
Lancashire PR1 1RE, UK

W. Van Puymbroeck
Commission of the European Communities
Rue de la Loi 200
B-1049 Brussels, Belgium

British Library Cataloguing in Publication Data Communications for manufacturing.
1. Manufacture. Applications of computer systems.
I. Withnell, S. (Stephen) *1951*– II. Puymbroeck, W. Van (Willy) *1955*–
III. Commission of the European Communities. Directorate-Generel for Telecommunications, Information Industries and Innovation
670.285

ISBN-13: 978-3-540-19642-6   e-ISBN-13: 978-1-4471-1820-6
DOI: 10.1007/978-1-4471-1820-6

Publication No. EUR 13015 of the Commission of the European Communities, Scientific and Technical Communication Unit, Directorate-General Telecommunications, Information Industries and Innovation, Luxembourg

© ECSC-EEC-EAEC, Brussels-Luxembourg 1990

Preface

Open Communications, essential for cost-effective Computer Integrated Manufacturing (CIM), was the subject of a congress sponsored by the Commission of the European Communities, DG XIII (Telecommunications, Information Industries and Innovation) at Stuttgart's Annual Machine Tool Fair. Experts from Europe, Japan and North America addressed the industrial needs for open systems for manufacturing and explained the technological solutions enabling a multi-vendor environment and full integration from design to manufacture.

The congress was modular to allow a selection of relevant sessions. The first day was aimed at policy makers who determine strategy and decide on technological investments. The second day was aimed at those responsible for applying the technology and needing an up-to-date technical understanding and awareness of future trends.

The work of ESPRIT project 2617, Communications Network for Manufacturing Applications (CNMA), was highlighted and explained. ESPRIT is an industry-driven, Community-funded research and development programme in information technology, and CNMA is a flagship project of the ESPRIT-CIM programme. The project is led by the British Aerospace and comprises users from the aerospace and automotive industries (Aeritalia, Aerospatiale, Magneti Marelli and Renault), vendors (Bull, GEC, Nixdorf, Olivetti, Robotiker and Siemens) and the universities of Porto and Stuttgart, together with the systems integrator Alcatel-TITN and the research body Fraunhofer IITB.

CNMA aims at furthering the development and adoption of industrial communications standards for CIM. It has commissioned several experimental and industrial pilot sites and gained a lot of practical experience on network architectures and on the use of OSI communication for the management, control and integration of manufacturing processes.

The Communications for Manufacturing congress was organised by the CNMA project team and CIM-Europe. CIM-Europe is an information and awareness activity of ESPRIT. Its function is to consolidate and enhance the effects of ESPRIT-CIM by disseminating information on progress and achievements of the programme. Further information on CIM-Europe's activities can be obtained from: CIM-Europe Secretariat, 200 rue de la Loi (office Breydel 9/54), B–1049 Brussels, Belgium.

Contents

Speakers' Organisations and Addresses

G. Blunck
Deere & Company
Deere Tech Services
John Deere Road
Moline, Illinois 61265, USA
Tel: +1 309 765 5199
Fax: +1 309 765 4128

N.E. Brownlow
BULL
1 Rue Carpeaux
Cedex 74
F-92039 Paris La Défense, France
Tel: +33 1 46 96 85 47
Fax: +33 1 46 96 80 92

S. Burgmeier
Kewill Systems Plc
Ashley House
20–32 Church Street
Walton-on-Thames KT12 2QS, UK
Tel: +44 932 248328
Fax: +44 932 254678

J.B. Cox
British Aerospace (Military Aircraft) Ltd.
Warton Aerodrome
Preston, Lancashire PR4 1AX, UK
Tel: +44 772 63 3333
Fax: +44 772 67 9193

K. Grund
EDS Deutschland GmbH
Central European SBU
Eisenstrasse 56
D-6090 Rüsselsheim, Germany
Tel: +49 6142 802347
Fax: +49 6142 802590

M. Honda
NEC Corporation
Basic Software Development Division
Daito Tamachi Building
14-22 Shibaura 4 – chome
Minato-Ku
Tokyo 108, Japan
Tel: +81 3 456 7431
Fax: +81 3 456 7497

A. Lederhofer
Siemens AG
AUT E514
Gunther-Scharowsky-Str. 1
Postfach 3220
D-8520 Erlangen, Germany
Tel: +44 9131 7 20879
Fax: +44 9131 7 33193

G. Krebser
ISW
Universität Stuttgart
Seidenstrasse 36
D-7000 Stuttgart, Germany
Tel: +49 711 2299234
Fax: +49 711 2299222

F. Marra
Syntax Factory Automation
Olivetti Information Services
Via Vela 27
I-10128 Torino, Italy
Tel: +39 11 5611888
Fax: +39 11 4360679

P. Martin
Centre UNIX BULL S.A.
1 Rue de Provence
B.P. 208
F-38432 Echirolles, France
Tel: +33 7639 7683
Fax: +33 7639 7600

G. Medici
Magneti Marelli S.p.A.
Innovazione Processi Produttivi e Logistica
Via Adriano 81
I-20128 Milano Crescenzago, Italy
Tel: +39 2 61830–255
Fax: +39 2 61830–206

J.J. Michel
CETIM
52 Avenue Félix Louat
F-Senlis, France
Tel: +33 44 58 34 40
Fax: +33 44 58 34 00

H. Portier
Aerospatiale
37 Bd. de Montmorency
F-75781 Paris Cedex 16, France
Tel: +33 1 42 242424
Fax: +33 1 42 242049

G. Segarra
Régie Nationale des Usines Renault
IGO6EGI Sce 0450
BP 103
F-92109 Boulogne Billancourt Cedex, France
Tel: +33 1 4609 6219
Fax: +33 1 4609 6330

T. Simmons
BAeCAM
Guild Centre
Lords Walk
Preston, Lancashire PR1 1RE, UK
Tel: +44 772 205164
Fax: +44 772 205241

J.M. Soto
Robotiker
Belako Elkartegia
E-48100 Mungia, Spain
Tel: +34 4 6740002
Fax: +34 4 6743273

R. Vio
FIAT
Corso Marconi 10/20
I-10125 Torino, Italy
Tel: +39 11 6565 422
Fax: +39 11 6565 707

K. Watson
Fraunhofer-Institute (IITB)
Fraunhofer Str. 1
D-7500 Karlsruhe 1, Germany
Tel: +49 721 6091 486
Fax: +49 721 6091 413

S. Withnell
BAeCAM
Guild Centre
Lords Walk
Preston, Lancashire PR1 1RE, UK
Tel: +44 772 205129
Fax: +44 772 205141

ACHIEVING CIM, USER PERSPECTIVES

CIM SYSTEMS COMMUNICATIONS
AT FIAT AUTOMOBILE DIVISION

Roberto VIO (*), Roberto BALDINI (#), Fulvio RUSINA'(+)

(*) FIAT Holding, C.so Marconi 10/20, 10125 Torino (Italy)
(#) FIAT Auto, Via Issiglio 63/A, 10141 Torino (Italy)
(+) SESAM, C.so Svizzera 185, 10149 Torino (Italy)

Summary

The increasing demand and competition in the manufacturing automation market are changing the approach to CIM implementations. Integration and therefore Factory Networking Systems are playing a more important role.

The aim of this paper is to present a management overview of the experience achieved, in the implementation of Factory Wide Communications, by FIAT Automobile Division (FIAT Auto) and planned evolution.

1. Introduction

The complexity of operations implied in Computer Integrated Manufacturing (CIM) projects is continuously increasing, due to the rate at which business and technological changes are taking place.

In order to gain competitive edge a manufacturing enterprise must define an integrated and heteroneous CIM architecture that will allow high quality productions at reasonable costs, reducing the lead time from product concept to marketplace delivery.

Viceversa a slow development of an efficient integration, that is computer-to-computer and computer-to-device communications, has inhibited the full evolution of CIM systems.

As an obvious consequence, interconnection/integration methods, and therefore communication solutions, must be developed between all components in order to define a complete CIM system that can be easily upgraded and reconfigured following market and competition demands.

The paper, after a summary on CIM architectures and communications in FIAT Auto (Chapter 2), points out, in Chapter 3, the experience gained by FIAT Auto in two projects; one in a production area, and the other in the area of engineering.

Finally, Chapter 4 describes planned evolution in the area of Factory Wide Communications, with emphasis on the most important FIAT Auto requirements.

2. CIM Architectures and communications

Several reference models for CIM architecture have been developed by the standardization committees, working groups and singular companies. Generally they have been realized with reference to constraints such as: simplicity, modularity and applicablility to a wide variety of industrial applications, taking into consideration also inter- connectivity problems, openess to new technology and independency from the computer and automation techniques in use nowadays.

Focusing on Manufacturing, one of the most referenced frames, is the ICAM (International Computer Aereospace Manufacturing) architectural model, where CIM activities are classified in 5 layers, each one identified by its functions, as shown below.

Table 1: ICAM Reference Model	
LAYERS	NAMES
Layer 4	Enterprise
Layer 3	Factory/Plant
Layer 2	Cell/Area
Layer 1	Controller/Device
Layer 0	Actuators/Machinery

Table 1: ICAM (International Computer Aerospace Manufacturing) Reference Model

At the base of the Model is Layer 0, which is concerned with the real prodution machinery, where the actual production takes place. Numerical Controllers, Robot Controllers, Programmable Logic Controllers, AGV Controllers, etc., are the devices managing the production, and are included in Layer 1 of the model.

Next, there is the Layer 2 (Cell/Area), at which groups of machine controllers are respectively managed or controlled and monitored. Factory wide operations such as: production management, scheduling, maintainance and quality controls take place in Layer 3. At the top (Layer 4) is the Enterprise level, where responsabilities lies for the Enterprise mission and a set of functions such as: financial planning, facilities planning and activities coordination with external bodies.

Focusing on Engineering, computer based architectures organized in four layers are generally taken into account. Enterprise, Engineering-Plant, Workgroup and Workstation are the general names of the layers. These are devoted respectively to: Enterprise computing, Engineering-Plant computing and storing, and computer-aided programming at Workgroup and Workstation level.

4

Communications

Communications are becoming more and more important in the definition of CIM computer architecture. Three main kinds of networks must be considered:

o   Enterprise Networks: the corporate data highway connects plants at different sites to the corporate headquarter through Wide Area Networks (WANs) or Metropolitan Area Netwoks (MANs). High-speed digital networks such as PSDNs/ISDNs (Packet Switched Data Networks / Integrated Service Data Networks), or high-bandwidth optical connectivity will become more and more common.

o   Manufacturing Plant and Engineering-Plant Networks: Typical communication functions such as: file transfer, message exchange, program-to-program, virtual terminal and resource sharing are used. Broad-or-baseband backbone Local Area Networks (LANs) are the solutions mainly considered.

o   Cell and Workgroup Networks: Focusing Manufacturing, due to the heterogeneity of the connected devices, ranging from simple sensors to complex machining centers and also performance requirements, this is one of the most critical areas. Generally devices are integrated by means of point-to-point connections or LANs. Concerning Engineering, due to the fact that only computer based systems are present, that is Workstations, LANs are the normal interconnection systems.

3. FIAT Auto Implementations

The first significant FIAT Auto experience in Factory Wide Communications dates back to 1985 when the concept of a main Factory Network was deployed for the automation of FIRE engine assembly in the Termoli plant. Later on in 1987, a number of large LANs were installed in FIAT Auto plants, covering different automation areas.

In the following paragraphs two implementations, in the areas of manufacturing and engineering are briefly described.

Cassino Manufacturing Plant

The CIM Architecture of the Cassino plant, as shown in Figure 1, is structured following the FIAT Auto CIM Reference Model, which is based on the ICAM frame.

In the area of networks, two different approaches have been implemented. A FIAT Auto communication standard, based on DECnet protocols as the factory backbone, that is used for communications between the various computers of the Cell and Plant levels. Different proprietary networks, implementing the cell networks, for the integration of device controllers and cell controllers.

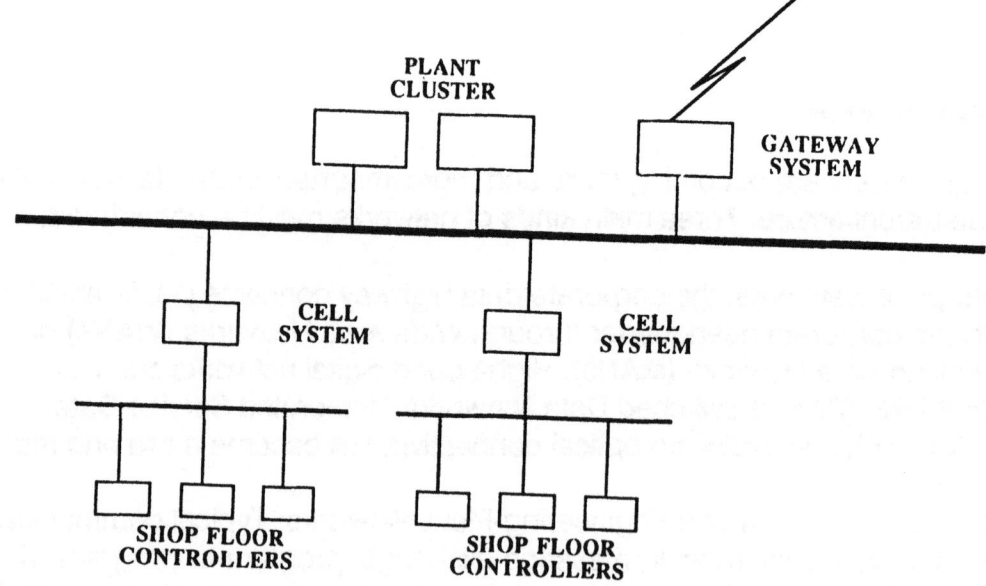

Figure 1: CIM Architecture of the Cassino Manufacturing Plant

FIAT Auto has developed a protocol named CABERnet to solve communication problems related to message synchronization, data delivery independence from receiver status, data integrity and restart features. CABERnet, considering the OSI reference model, is a protocol at the Application layer based on Digital DECnet, and is used to integrate the cell and the plant Controllers. The access methods and the media are respectively: Ethernet or CSMA/CD and IEEE 802.7 (Institute of Electrical and Electronics Engineers) broadband coaxial cable.

The proprietary networks such as Allen-Bradley Data-Highway, Texas TIway and COMAU HERMESlink are used to connect a large number of Programmable Logic Controllers, CNC Controllers and Robot Controllers to the cell controllers, named HERMES and developed by COMAU.

The resulting Cassino network architecture is impressive: more than 12 Km of broadband network, and about 1 square Kilometer of plane surface are totally covered by the network, about 150 nodes are directly connected to the factory network (VAX, MicroVAX and PDP systems, Personal Computers, Terminal Servers and SNA Gtw) and more that 1000
devices are connected to the total network (backbone and cell) system .

FIAT Auto Central Engineering

The Fiat Auto Product and Process Engineering Departments have in the last few years been developing a large Technical Office sistem devoted to activities such as: Computer-Aided Design (CAD), Engineering (CAE), and Design-Manufacturing (CAD-CAM).

The architecture, depicted in Figure 2, is based on a layered structure composed of four main layers, respectively: Enterprise, Engineering-Plant, Workgroup and Workstation. Systems at Workgroup and Workstation layers, are generally referred as computing "islands".

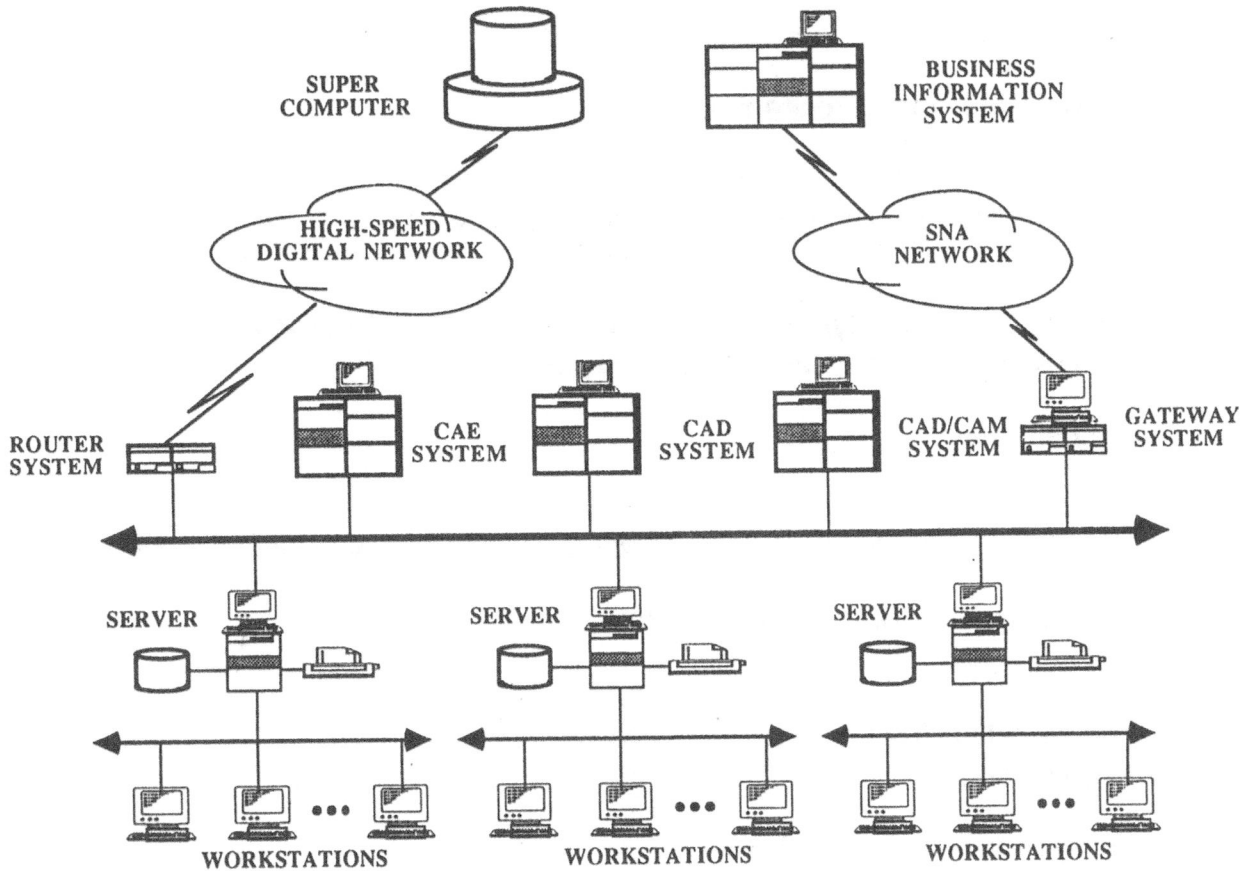

Figure 2: Computing Architecture of FIAT Auto Central Engineering

Enterprise and Engineering-Plant computing nodes, based on VAX systems and a Cray system, are mainly used for Engineering computation and as a central database store for the CAD systems.

The various islands, generally composed of a Server system and a number of Workstations, are mainly devoted to CAD development, CAE computation, and recently some of these have been devoted to the development of CAD-CAM programming.

All systems at the Engineering-Plant, Workgroup and Workstation layers are interconnected by means of a network based on baseband Ethernet, while high-speed digital networks are used to integrate computing systems at Enterprise Layer.

Due to the heterogeneity of systems installed, specifically, Cray and VAX systems at the Plant level, SUN and Computervision at the islands level, the Transmission Control Protocol/Internet Protocol (TCP/IP) has been chosen as the only possible solution, even if, in some case, some proprietary protocols such as DECnet are used.

4. Communication architecture evolution

For the future, FIAT Auto is looking for Factory Wide Communications solutions that will ensure higher interoperability among extremely heterogeneous devices and higher capability to cover the whole plant layout in a flexible way.

Optical Fiber, Open System Interconnection and Wide Area Networks are the strategic areas in which FIAT Auto is investing.

Optical Fiber

Optical Fiber technology is growing as an interest in the sector of industrial local area network, due to the multiple advantages they offer such as the followings:

o total immunity from elettro-magnetic and elettro-static noise,

o total isolation from electric noise, coming from power suppy lines,

o easy definition of complex/redundant topology,

o high bit-rate.

Recently a number of optical fiber standards have been designed in the IEEE 802 project. Particularly an optical fiber solution has been provided both for the three leading IEEE 802 standards: Ethernet, Token-Passing Bus and Token-Passing Ring.

One solution that is in the final stage of standardization is named Fiber Data Distributed Interface (FDDI). FDDI was designed by the American National Standards Institute (ANSI) by several dozen partecipating computer and telecommunications companies. FDDI uses a token-passing access scheme on a optical fiber media to achieve a speed of 100 Megabits per second (Mbps), which is an order of magnitude faster than Ethernet's speed of 10 Mbps.

FIAT Auto, in order to include fiber optics and in particular the FDDI in his industrial network strategy, is planning a migration path with the following milestones:

o to lay standardized optical fiber,

o to use optical fibers as a high-speed interconnection system for Ethernet networks,

o to define a full FDDI network as the backbone for an extensive industrial factory.

Open System Interconnection

The previous chapters have emphatized that two different solutions are nowadays widely used in FIAT Auto, respectively: proprietary, which uses many protocols such as DECnet, CABERnet and HERMESlink in the manufacturing environment, and a multivendor-one based on TCP/IP protocol in the engineering environment.

The final objective, in the area of Factory Wide Communications, for FIAT Auto is however an industrial network architecture, that will ensure an high interoperability among extremely heterogeneous devices.

The main prerequisite in order to achieve this objective is to define a network architecture based on Open Systems Interconnection (OSI). OSI is a data communication architecture, based on the ISO/OSI Reference Model, which allows for communication between systems conforming to the OSI standards. The OSI Reference Model has been developed by ISO (International Standard Organization) and is followed by the majority of information technolgy vendors.

Focusing on Application protocols, FIAT Auto is considering as candidate protocols for its OSI strategy, the most relevant protocols that have been standardized, or are in the final stages of standardization, such as:

o Manufacturing Message Specification (MMS) and MMS Companion Standard, as standard protocols at the Cell level,

o File Transfer Access and Management (FTAM), as standard file access system in multivendor environment,

o X.400, and X.500 as a standard solution for the Message Handling System and the Network Management and Directory Services in a multivendor enviroment,

o ISO/TP (Transaction Processing), as a protocol for the transaction processing environment.

The proprietary networking scheme, however, will continue to be selected in the future for some specific situations such as: i) when a single vendor is predominant in a certain area, ii) when exceptional performance is required, and iii) when it is not justifiable to remove existing networks.

Particular care will also be devoted to the migration of TCP/IP applications. Infact, even if TCP/IP offer a limited set of services such as: SMTP (Electronic Mail for ASCII text) ii) FTP (File Transfer for binary and ASCII) and iii) Telnet (Remote login to arbitrary hosts) it is however today the only networking solution to interconnect computers in a multivendor environment.

Wide Area Network

Emphasis will be placed in the future especially in Computer Integrated Enterprise (CIE) Technology. This means a lot of interest will grow in the area of Wide Area Networks.

Solutions such as Electronic Data Interchange (EDI), Odette (Organization for Data Exchange through Tele-Transmission in Europe) and the afore mentioned standard X.400, will play an important role in all the Wide Area Network solutions in FIAT Auto, both for inter-company and intra-company communications.

5. Conclusions

In summary, the lesson learnt by FIAT Auto in CIM Communication implementation indicates that Factory Wide use of de-facto or de-jure standard based networks always repays in terms of investment, flexibility, etc., while non standard compliant solutions always prove to be expensive and unflexible.

6. References

[1] R.VIO, "Planned Business Benefits of MAP Within FIAT", MAP-TOP-OSI Symphosium, Birmingham (GB), June 1988.

[2] R.BALDINI,"Cassino: Bradband Local Area Network System", FIAT Internal Technical Report 1987.

[3] M.JOHANSSON, "Communications Issues in Manufacturing", CIM Review, Fall 1989, Auerbach Publishers New York (NY).

[4] L.M.OLIVA, "Why Standards are so Important to CIM ?", AUTOFACT '89 Conference Proceedings, November 1989, Detroit (MI).

[5] Open System Architecture for CIM, ESPRIT Consortium AMICE, Springer-Verlag Edition 1989.

[6] M.BOBBIO, F.RUSINA', "An Approach to Factory Control Architecture Using Distributed Cell Controllers", 19th ISATA International Symposium on Automotive Technology and Automation, October 1988, Monte Carlo (MC).

BUILDING THE CIM , NOT AN OTHER BABEL TOWER.

G.Segarra

Régie Nationale des Usines
Renault , IGO-EGI Sce 0450
BP 103 , 92109 Boulogne
Billancourt Cedex FRANCE

Summary

Facing the rapidly growing complexity of manufacturing systems , RENAULT started a CIM programme aiming at specifying an integrating infrastructure allowing to limit strongly the manufacturing system complexity and to satisfy the company objectives of TOTAL QUALITY , FLEXIBILITY , PRODUCTIVITY and COST OPTIMIZATION. This programme is essentially based on two major activities which are STANDARDIZATION and RESEARCH .The standardization activity consisted to select standard profiles and services applicable in the manufacturing area . After an evaluation phase RENAULT selected UNIX and MAP 3.0. The research activity consists to defining an Integrating Infrastructure , of course comprising communication standards, but also including the definition of integrated standard data formats , application constructs (building blocks) and high level interfaces . This research activity is undertaken in relation with the European ESPRIT programme and includes cooperations in CNMA , DELTA 4 and VOICE projects such as described in this paper .

1. Company objectives

Three major objectives of RENAULT are :

- Total quality,

- Flexibility ,

- Productivity and cost optimization.

Of course these objectives apply to the factory production process in which the information technology and telecommunication are coming more and more into prominance . At the level of factory data processing systems , these company objectives can be expressed in the following terms:

11

1.1 Total quality

The total quality of the production system can be obtained through the following actions :

a) Keeping the capability of constantly mastering the overall system in spite of a rapidly growing complexity. That is to say keeping the control of the overall system whatever the evolutions necessary to be undertaken to adapt it to new needs and new technologies emerging during its life cycle .

b) Maintaining a high level of availability of the overall system (ZERO FAULT perceived by the manufacturing process) . Such important goal can be reached by both preventive maintenance actions and by introducing some redundancy in the system in order to reconfigure it on the event of a sudden failure .

c) Permanently checking that the obtained system response time stays within predefined limits such as required by applications and this whatever the workload evolution of networks and systems .

d) Insuring the information consistancy (even during systems failures) and offering to the end users and the manufacturing system an information presentation form adapted to their needs and jobs .

1.2 Flexibility

1.2.1 system flexibility
The flexibility of a system can be defined as its capability to evolve easily (with a minimum of effort) in order to integrate the changes consecutive to:

- Architecture evolutions , allowing various degrees of hardware/software components distribution relatively to systems cost/capabilities evolutions or to the availability of new network/systems architectures .

- Technology and concepts evolutions , allowing the integration of new technologies (ie : fiber optics , multi media , neuronal systems , ... etc) and concepts (artificial intelligence , object oriented concepts ,... etc) .

- Application evolutions , allowing to automatize new subsets of the production process and integrating the new applications resulting from this automatization action with the existing ones.

- Scale evolutions , allowing to extend in size the spanning and coverage of some existing applications .

In fact the perennity and cost of the system will be directly related to its flexibility, since a flexible system will easily evolve thus avoiding its complete rebuilding due to the impossibility to adapt it .

1.2.2 management flexibility

Another important flexibility aspect is that the company management must keep its complete choice freedom when buying new EDP and information systems. This means that all vendors ' products get the same chance to be retained relatively to the installed base and that the choice be not constrained by the preeminence of some vendor proprietary existing solution. Such flexibility can be obtained by selecting only OPEN SOLUTIONS based on international standards. An open solution approach will allow to develop a flexible purchase policy limiting the risks related to:

- Vendors'product strategy evolutions,

- Vendors' company failures or take over,

- Vendors' product family limitations and shortages,

- Vendors' quality degradation,

- External constraints placed by the EEC and governments... etc.

1.3 Productivity and cost optimization

The productivity of a company will be increased in some cases by some automatization actions but also primarily by improving the information circulation between operators (HUMAN and MECHANICAL) themself and with the factory information system. Facilitating the information flow, will lead to the funtional integration of the overall system (CIM objective). Such integration could become extremely difficult and a very costly experience if before, some optimization actions are not undertaken to simplify the overall information system . Consequently it is a strong goal to take the opportunity of this systems integration action, to simplify and then optimize the overal system. Such action then will impact the total cost in the following manner :

1.3.1 hardware cost
Two kinds of hardware costs have to be considered :

- Network cost,

- Systems cost.

a) The network cost can be optimized by cabling only one time the factory and by limiting the number of LANs and protocoles stacks options. In order to reach this goal it becomes necessary to be very carefull when selecting LAN technologies and protocoles, retaining only those which offer enough capabilities to satisfy applications requirements for a long period of time (ie:10 years perennity) .

b) Systems cost can be optimized by sharing some common resources (files servers , data bases , communication servers , editing servers , ... etc) and by selecting data processing equipment well adapted to the applications requirements .

Moreover , a flexible purchase policy based on open systems will allow to select the best solution satisfying the users' requirements , and this , after a selection process allowing vendors to compete . The fact that several offers are in competition encourages vendors to make their best offer , often leading to a significant cost reduction (10 to 30 % discount) . It has to be noted also that , as open systems are being more and more used , the mass effect produced by the market increase will accelerate the vendors' return on investments leading to a gradual cost decrease as the market grows up .

1.3.2 application software cost
Software development and maintenance costs are rapidly increasing with the development of the tremendous hardware capabilities which are enabling the building of powerfull applications . Consequently it becomes urgent to develop modular , reusable pieces of software (building blocks) portable in various operating systems ' environment and application contexts .

1.3.3 other costs
This includes :

- The multiple competences necessary to manage , design , install , operate, maintain complex systems . Optimization of such cost can be obtained by reducing the complexity of the overall system .

- Documentation ,administration /maintenance tools , integrating infrastructure ... etc . Once again cost optimization will be possible through the utilization of a minimum number standard components .

2. The growing complexity

Currently several factors are contributing to the complexity increase of the overall production system , they are :

a) The proliferation of manufacturing devices and data processing systems at all levels of the factory (currently more than one thousand with an annual growing rate of 10 to 20%) .

-b) The heterogeneity of these devices and systems coming from more than 15 different vendors :

. Manufacturing devices such as PLCs, robots , machine tool ,measurement machines , AGVs ,... etc;

. Video systems such as vision systems , workers training systems , ... etc.

. Data processing systems including CAD workstations, industrial micro/mini computers , Dumb terminals , mainframes ,PCs ,... etc.

c) The necessity to interconnect more and more frequently these devices/systems to form an integrated system through several levels of local area networks .

d) The various application requirements in terms of network instantaneous/average bandwidth , network/systems dependability, response time ,... etc .

e) The large distribution of these devices/systems over areas reaching 1 million of square meters and the harsh manufacturing environment (EMI, vibrations , heat , dust ,chemical polution ... etc).

f) The variety of manufactured products options and the complexity of the manufacturing process itself which evolve towards new manufacturing technics such as just in time , electronic KAMBAN, modular manufacturing ... etc.

The only way of limitating and mastering such complex system is to have a global integrating approach based on a limited number of standards . Some of these standards are currently available (ie : MAP 3.0 /TOP 3.0 , CNMA 4.0 .. etc) but other are strongly missing . It becomes then a primary role of the research to contribute to the development of the missing CIM standards .

3. The research and standardization winning combination

As already mentionned a certain number of standards are emerging from the International Standard Organization (ISO) and European standard organization (CEN/CENELEC/ETSI). These standards are contributing to the Enterprise modelization (CEN/CENELEC/ETSI ENV40003 (ref 2)) or to the building of an open integrating infrastructure (basic OSI standards used in MAP 3.0(ref 1) and CNMA 4.0 (ref 3) . But all these standards represent only a small part of what is necessary to build the CIM with standard building blocks coming from different sources (users , vendors , engineering companies ... etc) . From a user point of view , the conditions necessary to meet to build the CIM are the following :

a) Getting the possibility to buy on the market , products complying to a standard Integrating Infrastructure including :
. Open communication systems well adapted to the factory environment and CIM applications context .

. Open distributed operating systems offering a stable standardized support environment for CIM applications . That is to say , offering high level application interfaces, standard programmation environment and languages.

. Standard front end servers such as data bases or communication servers allowing an open access to common ,sharable data and communication services through standard local area networks .

b) Having a real possibility to develop a "make or buy" policy , that is to say for the company to decide by itself to develop or just buy the necessary functional building blocks on the basis of strategical , economical or resources availability criteria. In order to get such possibility it is important to standardize open application constructs (generic, partial) from which will be instantiated particular application software components adapted to the company requirements .

We think that these conditions can be met by research actions leading to experimental development combined with contributions to the standardization process.

4. The Renault MOSAIC programme

MOSAIC (**M**aitrise et **O**ptimisation des **S**ystmes et **A**rchitectures **I**nformatiques **C**omplexes) meaning mastering and optimizing complex data processing systems and architectures , is a research programme aiming at completely specifying and testing a CIM Integrating Infrastructure for applying it within RENAULT industrial systems . This research programme is undertaken in strong liaison with the European ESPRIT (**E**uropean **S**trategic **P**roject **R**esearch in **I**nformation **T**echnology) programme and includes contributions in standard organizations . The cooperation with ESPRIT is achieved through the following complementary projects (fig 1):

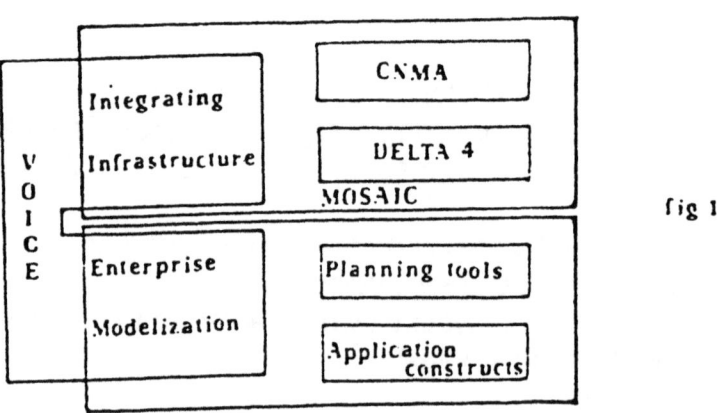

fig 1

16

- ESPRIT VOICE (Validating OSA in Industrial CIM Environment).

- ESPRIT CNMA (Communication Network for Manufacturing Applications).

- ESPRIT DELTA 4 (Definition and Design of an open Dependable Distributed system architecture)

4.1 VOICE project

The objectives of the voice project are :

- To demonstrate that the open system architecture proposed by ESPRIT CIM-OSA (AMICE consortium) is a suitable framework for integration and overall manufacturing system optimization .

- To contribute to the rapid proliferation of CIM components within European industry .

- To contribute to the overall CIM-OSA standardization effort .

In that respect , RENAULT will concentrate on :

a) The definition and application of front end services of the integrating infrastructure, taking profit of the CNMA and DELTA 4 projects results here under described .

b) The macroscopic modelization of a factory and microscopic modelization of cells contained within the macroscopic model. This modelization action will be achieved for the function and information views.

c) The definition of integrating infrastructure and application constructs used to validate the CIM-OSA concepts . Some generic and partial constructs will be specified and implemented in a particular manner in order to satisfy some company requirements .

This ESPRIT project is a new project supported by the EEC and regrouping the following partners:

- IPK FRAUNHOFER, Germany
- ISMCM, France
- ITP-TNO, Netherlands
- KFK, Germany
- PRISME/ZENON, Greece
- RENAULT AUTOMATION, France
- RENAULT AUTOMOBILE, France (prime contractor)
- TRAUB AG, Germany

4.2 CNMA project

CNMA phase 4 is reaching its final demonstration stage and will continu in phase 5 starting beginning of 1991. The results of the phase 4 will be used as a basis to continu the work in phase 5 . The major achievements of phase 4 are related to :

- MMS

- MMS Application interface

- Network Management .

An integrated network management system (ref 4)prototype will be demonstrated . This management system based on emerging network management ISO standards (CMIS-CMIP , SMFA ....) contains important tools used for different activities of the industrial network life cycle (fig 2) . The five activities supported by CNMA network management system are :

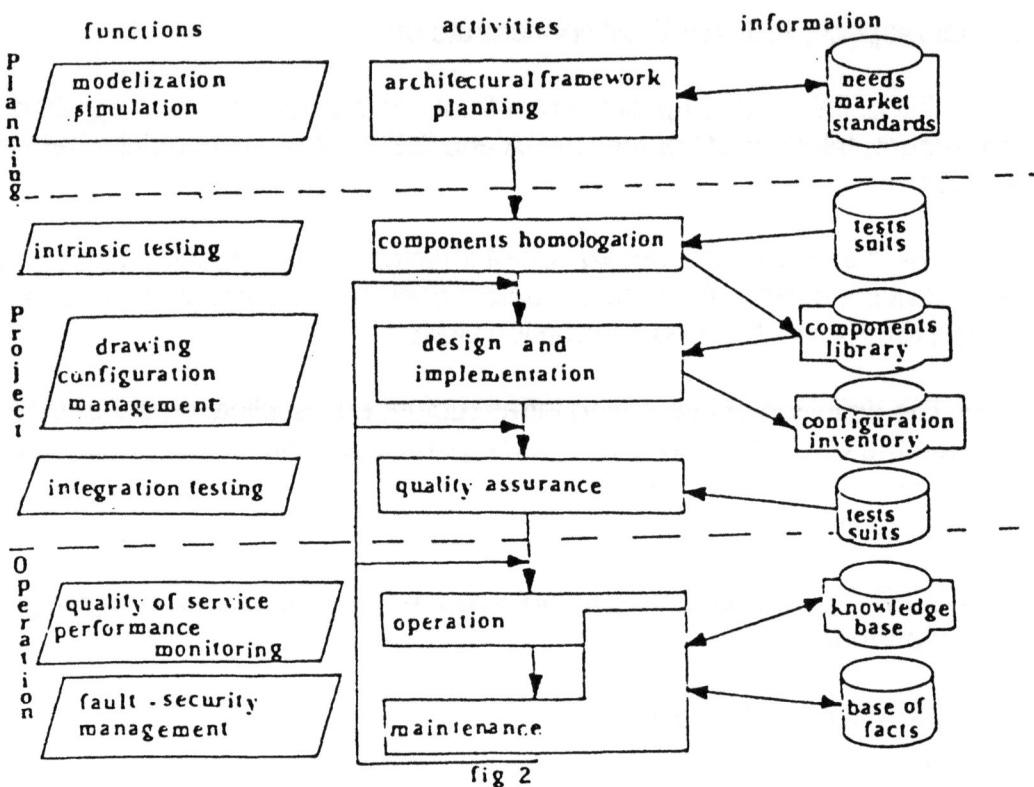

fig 2

a) network design

The network management console(an X terminal) will be used to draw a complete extended LAN topology comprising network hardware components (cable segments , taps , repeaters , amplifiers , remodulators .... etc) and systems (intermediate / end systems) hardware/software components . The topology will be built by instantiating homologated components contained and described in a components library. The components description will include several representations:

- Component characteristics and associated attributes (information view).

- Component physical desciption (physical resource view).

- Component functional description (functional view).

- Icon representing the component .

b) network/systems configuration

This activity consists to configure network/systems manageable components , checking their object attributes values and possibly up dating them (replacing their default value) when necessary .

c) network/systems quality assurance

Several tests will be supported by the network management system allowing to assess the quality of the network and installed systems . These tests include :

- Connectivity testing ,

- Data communication tests,

- Connection/data saturation tests .

From the network management system it will be possible to specify the configuration to be tested (group of stations and their relationships) and the test sequence to be executed .

d) network/systems monitoring

During the operation of the overall system , the network management system is constantly monitoring all manageable network/systems objects . This monitoring is achieved in order to assess the level of service quality / performance offered by the communication network . Instantaneous trends and long term statistics will be obtained repectively for network/systems maintenance and planning .

e) network/systems maintenance

The maintenance activity comprises the up dating of the overall system taking into account architecture /applications evolutions , but also essentially fault management activities allowing to achieve preventive maintenance on trends analysis and curative maintenance on problem detection and analysis . The overall communication network maintenance will be aided by the knowledge based system KOOL running under UNIX and accessing the management information base distributed in the network manager and agents .

4.3 Delta 4 project

Delta 4 project (ref 5)is complementary to CNMA in that sense that it concentrates on distributed operating systems and applications running in a fault tolerant environment . Both projects CNMA and DELTA 4 have common partners (BULL , IITB FRAUNHOFER , RENAULT) facilitating the harmonization of some aspects such as the network/distributed system administration . Consequently there are some good chance that the management tools being used by CNMA be extended to cover distributed operating systems and applications management .

Fundamental to DELTA 4 are the three following concepts:

a) A dependable and therefore fault tolerant communication system called MCS (**M**ultipoint **C**ommunication **S**ystem) offering reliable , atomic multicast communication over standards LANs such as ISO 8802.4 , ISO 8802.5 , FDDI .

b) An abstract high level application support environment (DELTASE) providing to the applications programmer an homogeneous environment in which the underlying hardware and support software complexity is masked . Thus the programmer views a set of interacting objects which after generation and installation operations are transformed into run time software components which can be distributed and replicated relatively to some system management strategies taking into account the following aspects :

. Applications criticity,

. Hardware failure modes,

. Resources optimization (load balancing),

. Performance and availability constraints.

The oriented object concepts evaluated in DELTA 4 are strongly in line with the emerging ODP (**O**pen **D**istributed **P**rocessing) ECMA/ISO standard .

c) A distributed administration system allowing to configure the overall system accordingly to the user's reconfiguration strategy deduced from the here above identified aspects .

4.4 The MOSAIC Programme

The MOSAIC programme will then contribute to the definition and evaluation of an Integrating Infrastructure allowing the building of consistant , optimized and modular fault tolerant distributed systems comprising:

- Object oriented concepts and tools allowing the building of applications constructs independent of the underlying operating system environment .

- High level application interfaces allowing to mask to the programmers the physical location and redundancy level of interacting objects .

- Standard communication systems integrating existing profiles based on ISO such as MAP 3.0 / CNMA 4.0 with new multipoint profiles adding important services such the reliable /atomic multicasting of messages .

- Systems/Networks administration tools allowing to master the evolutions of the overall system including the application constructs from a small number of supervisory points .

Experimentation of this Integrating Infrastructure will be achieved through the implementation of CIM applications taking place in pilot installations such as proposed within the scope of the ESPRIT VOICE , CNMA and DELTA 4 projects .

5. Conclusion

We tried to show through this article that there is not any competition between research and standardization and that in the contrary these are two complementary processes . The research feeds the standardization process with new solutions aiming at satisfying users needs which are not yet covered by the current standards whereas the standardization allows to capitalize on past research action results thus avoiding to rediscover constantly the wheel. Such approach is of course valid in the case of MAP which is an important capitalization step in the area of heterogeneous systems communication and which is now enabling the users to start the building of CIM systems . However MAP 3.0 do not satisfy all users requirements and must be complemented in some areas such as:

- Fault tolerant systems,

- Network administration ,

- Remote data base access ... etc.

Such extensions then can be added in an upward compatible manner keeping existing LAN standards and some major application services like MMS/FTAM and identifying an integration path allowing the transparent cohabitation within end systems of different profiles satisfying different needs .

It has to be noticed that the essential users investments are in the factory cabling system and in application software development . Consequently this investment must be protected leaving the possibility of basic operating / communication software evolutions provided that the installed LANs and application software be reusable. Standard high level application interfaces are major building blocks enabling application software portability in various operating and communication system environments .

References

(1) Manufacturing Automation Protocole specification version 3.0 . European MAP Users Group.

(2) ENV 40003 : Computer Integrated Manufacturing , System Architecture , Framework for Enterprise Modelling . CEN/CENELEC may 1990 .

(3) CNMA Implementation Guide 4.0 ,volume 1 and addendum 1 , 2617 T profile ESPRIT project 2617.

(4) The network management aspects in ESPRIT 2 CNMA project by F.LANGLOIS and G.SEGARRA . ESPRIT WEEK proceedings 1989 .

(5) DELTA 4 Overall system specification 1988, Edited by D.POWELL .

AEROSPATIALE

DEVELOPMENT OF CIM TOOLS

H. Portier

AEROSPATIALE
37, Bd. de Montmorency
75781 PARIS CEDEX 16
FRANCE

1. Introduction

In a fast changing economical world, the enterprise environment which used to be stable and where forecasts could be made, is now moving at random in other words dangerous. The consequence is a selection process in which only companies flexible enough to match the competition at world level will survive. Controlling the main parameters which typify the enterprise reactivity is thus a fundamental challenge and CIM is one of the answers, not only for survival but also for increasing the market share and ensuring growth. In french language this concept is know as "Productique".

2. "Productique" : definition, scope

Despite many publications and related studies this french word is still used with very different meanings. Its common root with "production" and "productivity" has often limited its field to workshop, ignoring design or support activities. In order to define the various fields covered and to clarify the context of this presentation we propose our own definition and a quick look at all technologies involved.

Productique : concept and framework used to improve simultaneously productivity, quality and flexibility of an industrial entity by means of information technology and automation.

3. Strategy

Based first on an integration as complete as possible of all enterprise activities with common definition of data and processes and full automation of their creation, management and use. Basic tools here are information technology and automation, applied to all enterprise acitivities. CAD/CAM, Robots, Numerically Controlled machines are just illustrative examples, the ultimate result of integration being the complete deletion of frontiers within the enterprise.

But implementing CIM simultaneously in all departments is very difficult to plan : priorities, budgets limits, staff capacity have to be taken into account and a methodology is needed. This is why, for discrete objects manufacturing industries, an approach based on a matrix is proposed, with group technology providing the families listed along the horizontal axis.

Complete application chains from preliminary design to support can then be isolated and integrated : sheet metal parts, turned parts, 2,5 D milled parts, composite materials parts, electrical harnesses, assembly operations... etc... Integrations are achieved separately for each technological family "vertically" in the matrix but in keeping reasonable homogeneity along each horizontal line, i.e. for a given type of application (ex. process planning).

4. Integration and the enterprise environment

The very large number of problems to be solved in order to reach integration of the applications chains in a given enterprise results in the involvement of many external suppliers and emphasizes the need for standards agreed by vendors of machine tools, A.G.V., computers, LANS... etc.

Several bodies are confronted with such a challenge and are working in many different ways to improve the situation : ISO, CEI, CEN/CENELEC (ITAEGM), ESPRIT, EUREKA, ODETTE, AECMA, CAM-I... etc..The ISO technical committee 184 is the key player. At european level CEN/CENELEC has set up a specialised working group (Information Technology Advisory Expert Group on Manufacturing, know as ITAEGM) whose workprogramme has now been approved.

In France, the Ministry of Industry and AFNOR are using the CONEP group (Comité d'Orientation de la Normalisation Européenne en Productique) to coordinate actions at national level.

5. The enterprise information system

The most important system in the enterprise it is also the most difficult to integrate. The elements to be taken into account, go from the physical media providing the link to the man/system interface, going through transmission protocols, data base management systems, operating systems...etc.

It is probably necessary to state again here the conditions to be met to achieve a true CIM system :

- automation an/or computer assistance for all main enterprise activities (design, production, quality assurance, support...etc).

- automated flow of informations and objects (raw materials, parts, components, end-products)

- activities and flows automated in order to achieve the most flexible enterprise response to market needs and to optimize productivity

Under pressure of I.T. vendors the problem has often been reduced to the famous OSI "7 layers model" primarily intended for file transfer between computers, with most obstacles located in layer 7 (called "applications layer"). It can be seen that the seven OSI layers hardly cover 25 % of the total integration problem. It is also an apportunity to position several well known acronyms : MAP, TOP, FIP, IGES, SET, VDA/FS, STEP, GKS, PHIGS...etc.

6. An example of integration tool : S.E.T. (Standard d'Echange et de Transfert)

One of the key fields in systems integration is the management and storage of data. Whatever the enterprise activities the assumption should be made that no vendor can provide the complete set of tools needed for automation of preliminary design, detailed studies, production, quality assurance, testing or operation support. Adding the industrial risk inherent to a unique source and the benefits of competition between vendors leads to a basic problem : storage and selective retrieval of data coming from various non compatible systems. The only well know answer is a neutral language, minimizing the number of required interfaces.

An old example -more than ten years old- but still not well solved is the case of exchange of informations between CAD/CAM systems. Such a neutral language has not only to cover exchange of common entities but also to allow archiving of 100 % of data produced by any CAD/CAM system and even by systems to come on the market in the future, with new types of entities not yet defined.
In the seventies the only solution consisted in developing specialised interfaces.
In 1979 IGES (Initial Graphics Exchange Specification) was launched in the U.S.A.

Dealing only with exchange of data (no archiving capacity), using old fashioned information technology (card image) and contrary to the CAD/CAM vendors commercial policy this project has never produced an industrially usable set of interfaces. In 1984, the International Standards Organization took the lead with the so-called STEP project, again controlled by the U.S.A (ANSI/NBS). Today this project his lagging behing objectives, the most optimistic forecast beeing an availability of a first Draft Standard in 1989 and many fundamental questions are still awaiting answers.
In this environment, Aerospatiale, confronted with this problem at the end of the seventies and realising the failure of IGES at the beginning of the eighties, took the decision in early 1983 to proceed with the development of its own standard : SET (Standard d'Echange et de Transfert).

After development and implementation of several interfaces (Computervision, Cadam, Anvil...etc.) for internal use the Airbus A320 project gave an opportunity to score a world premiere : the exchange by four major partners of an international project (British Aerospace, CASA, MBB and AEROSPATIALE) of all drawings needed to enable the separately designed and build subassemblies (wings, fuselage sections, tailplanes, engine pylons and nacelles) to match on the Toulouse assembly line.

This success story had consequences : the adoption of SET interfaces for the development of the european space shuttle Hermes, for the Ariane 5 launcher and recently, for the Airbus A340 project.

Industrial tool from the beginning, SET became a french AFNOR standard in 1985 (Z 68300) and is now supported by other bodies : the automotive industry and the Ministry of Defence.

The work done in common by Aerospatiale and these bodies is proposed as french contribution to ISO/TCI84/SC4 for the development of STEP.

7. European R&D programs in CIM

Both ESPRIT -European Community initiative- and EUREKA a framework sponsered directly by the European governments have CIM projects.

Within ESPRIT on of the most interesting ones is CIM/Open System Architecture. After nearly three years this project is nearing completion of a proposed architecture for CIM systems which should make integration problems a lot easier in the future.

Under the EUREKA label five european aerospace companies (Aeritalia, Aerospatiale, British Aerospace, CASA and MBB) have teamed to propose EUROPARI, a group of CIM projects.

EUROPARI is proposed as an Umbrella project for providing a framework for the development of Advanced Manufacturing/Computer Integrated Manufacturing Technologies within a five year timespan. Four specific areas have so for been identified within EUROPARI :

ECRAS   - Advanced Composite Manufacture
EIFAS   - Assembly Systems
SPACE - Electrical Assembly/Installation
SPIDER - Small Metallic Parts production

A close association is also defined within EUROPARI with :

PARADI   - Production Management in highly automated factories.
                  (EUREKA) labelled in 1986, phase 1 completed)

The primary drive behing these project areas is collaboration between the partners and the cost effective utilisation of their respective limited resources to achieve a reduction in development timescales.

The technological drive is to develop specific tools by way of an integrated approach to solve manufacturing problems. These will encompass integration right through from design, Production Engineering, Manufacturing to product support ; it will involve, for example, IKBS (Intelligent Knowledge Based Systems) and data management.

The five projects are :

ECRAS (European Composite Reconfigurable Automated System)

The flexible integrated automated system for the design and manufacture of fibre composite materials.
Modern manufacturing is making increasing use of new composite material technologies for example Carbon Fibre Composites and a great deal has been learned in the development and design and manufacturing processes of this emerging technology.
Although a high technology end product much of the manufacturing routes and quality assurance techniques are prolonged and labour intensive.

It is essential that industry, together with research activity to ensure both the product consistency and material integrity emphasised the growth of the industrialisation of the productive process.

The main factors influencing large scale application of these materials are :

- Production Cost
- Production Quality (réf. design and production processes)
- Material integrity (réf. environmental degradation)

EIFAS (European Integrated Flexible Assembly System)

The development of an integrated flexible assembly system.

The problems relating to the assembly of complex products are characterised by :

- a) large inventories due to high degree of completion and hence value, of the assembly items

- b) a wide range of disruptive factors, since the full complexity of the total product must be considered during the assembly process and the probability that parts are missing, for example, becomes relatively high

- c) considerable risk to the schedule since only little time remains in the programme to provide a buffer for disturbances during assembly.

The primary drive within EIFAS has been to look for systems to cover the assembly process from the design to tinish product stage. The prime considerable of the projects in the areas of :

- Product Design
- Assembly System Design
- Assembly Production Planning
- Assembly Production Control
- Logistics
It is required to develop an integrated CIM approach with the appropriate Manufacturing Technologies to address these types and problems to enable efficiency inrease and to improve profitability ad performance.

SPACE (Système de Production Automatisé de Câblages Electrique)

The automation of electrical harness assembly prior to installation to aircraft.

Electrical assembly is labour intensive with a high degree of change and configuration management. Within Space, a wide range of activity areas are proposed, which cover the design and integrated assembly of boxes and panels prior to assembly to manufacture of wiring harnesses.

The final equipping of airframes is a highly complex demanding task and involves amalgamation of harnesses and installation of components and sub-assemblies. This is also addressed within the project brief to enable the effective application of technology.

Testing of electrical systems is also of paramount importance both in development and final assembly during production. The existing processes are in themselves highly complex and sophisticated but could be integrated with the rest of the design/production process to achieve a high degree of the vertical integration.

SPIDER (Système de Production Intégré d'Eléments Regroupés)

This project area is aimed at developing an integrated flexible system to produce "simple" metallic parts. That is for both machined flat and formed, pipework. The scope, like the other projects, covers the development and integration of the vertical activities associated with small metallic parts production. Large numbers of detail parts are manufactured in small batches for each project. With end-product diversity being increasingly demanded by customers and more rapid turn round times on new variants or modifications it is important that manufacturing companies can keep pace. The profitability of such a manufacturing operation has to be developed hence the tools to enable these activities to be undertaken have to be developed.

All components demand process planning tooling, raw material, machine operators, production control, etc. to expedite work through the manufacturing cycle. An interface with Design, Stress and other services or upstream Departements is also required. During the high degree of change undertaken during the manufacturing life of a project the degree of complexity of the change management process is considered.

Increase in control and reduction in reaction and elapsed, during the manufacturing process, are the prime drivers towards the instigation of SPIDER which will examine sheet metal, prismatic machined parts, rotational (turned) parts and pipes, again in the sense of a vertical integration plan.

PARADI

PARADI has EUREKA status, as a separate project already labelled to include this project within the umbrella of EUROPARI.

Its main objective is to optimise overall production management, taking into account integration and automationof the vertical integration functions, i.e. from preliminary disign through to end product support for the various technologies involved within EUROPARI. The various aspects involved with the specific technology areas must be considered within EUROPARI, as PARADI is designed to take a global view across a broad range of activity. Essentially Paradi represents a horizontal integration activity across the four vertical activities of ECRAS, EIFAS, SPACE and SPIDER.
It is seen that this approach is consistent with an effective CIM implementation and provides for a coherent approach avoiding duplication of effort.

The EUROPARI management committee, via a working group, is identifying and defining interfaces between PARADI and ECRAS, EIFAS, SPACE and SPIDER. Common pilot sites for Paradis and one or several other Europari projects are being identified in order to make sure that integration between Europari vertical and horizontal projects is a reality.

8. Conclusions

A few comments :

- Knowing international environment is an essential prerequisite in CIM : methods, tools, systems which are becoming "de jure" or "de facto" standards, integration limits that can be achieved to day or tomorrow, etc. All these parameters govern investment and development effort.
- Sophistication of systems integration often needs in a modular approach, adapted to the state of the art in each activity and to the investment capability. Nevertheless such an approach must be organized within a general framework, approved at enterprise top level.
- No vendor, whatever its size or quality can deliver a CIM turnkey system for a complete enterprise. And even if all required modules are listed in its catalog, competition between suppliers is a must for a company aiming at the best possible performance.
- Standards are often used to support sales and help building integrated systems. But only performances, delivery dates and costs guaranteed by contract must be considered even if offers involve well advertised standards.
- Environment knowledge, best suppliers and standards are not enough to guarantee success : the most difficult step is towards integration of people and teams. There are examples of low integration heterogeneous systems working properly in the hands of motivated teams and -at contrario- of expensive "state of the art fully integrated CIM systems" running far below expected performances. This is where we leave CIM for a much more complex field -man behaviour- and where I have to stop speaking, by mere lack of knowledge.

PRODUCTION AUTOMATION AND INTEGRATION

PERSPECTIVES IN SMALL AND MEDIUM ENTERPRISES

J.J. MICHEL

CETIM (Centre Technique des Industries Mécaniques)
52, avenue Félix Louat - SENLIS - FRANCE

Abstract

Save communications problems in order to increase productivity is the major present goal for every kind of enterprises. Existing equipments in small and medium enterprises are often completely heterogeneous as in large ones and provide the same problems, however the significance of such a situation is for them much higher. They don't have generally sufficient assets to launch into "ex nihilo" design and realization of whole automated production systems, including I*.LAN's. They often are legitimately afraid by the cost of operation, that can, in addition, greatly disturb their current business. On the other hand supply is, now, very large, complex and sketchy. Facing this situation, there is (in small and medium enterprises) a lack of both human ressources and ability to master the whole complexity and variety of the matter. One solution will, perhaps, be the improvement of engineering companies that would help enterprises to build their own applications. Engineering companies will, thus, help SME to design each part of their own communication system,and in the long term use standardized components in order to ensure a good garantee of a long service life, compatibility with new evolutions and global perspective.

1. INTRODUCTION

For many years (after a large period during which the automatisation of manufacturing task has been increasing rapidly), companies of any size are more and more faced with the problem of automation island's multiplication and with the lack of communication between equipments which are unavoidably heterogeneous.

In 1980 General Motors had already observed that about 50% of its automation's applications cost was due to solving communications problems. Therefore, the company started the project which later will provide MAP (Manufacturing Automation Protocol)[1], based on international standards (if available). General Motors was rapidly joined by several of the major world's companies which were gathered together in regional and world MAP user's group : MAP/TOP world federation, Australian MAP/TOP Interest Group (AMIG), European Map Users Group (EMUG), Japanese MAP User's Group (JMUG), North American MAP/TOP User's Group (NAMTUG). Recently the East European MAP/TOP Interest Group (EEMIG) was added to the first ones.

At the same time, over several years, a more comprehensive line of thinking was developed both on communication architecture enterprise modelling, integration problems, and on standardization in this field. This high activity led recently in Europe to some major actions like CNMA and CIM-OSA Esprit project. This also provided some major documents from the European Standardization Organization : M.IT.04 (inventory and taxonomy of all standardization work to be done or already in progress, with a description of problems to be solved and priorities indications)[2], the European prestandard ENV 40003 [3],recently adopted, suggests a framework for enterprise modelling greatly based on CIM/OSA Esprit project results.

Now, in every country, major user's companies are putting significant resources on CIM (Computer Integrated Manufacturing) studies, evaluations, developments and realizations. Thus the CIM market (not only restricted to the communications and I*LAN's market) has grown rapidly throughout the last few years as shown in Figure 1 [4].

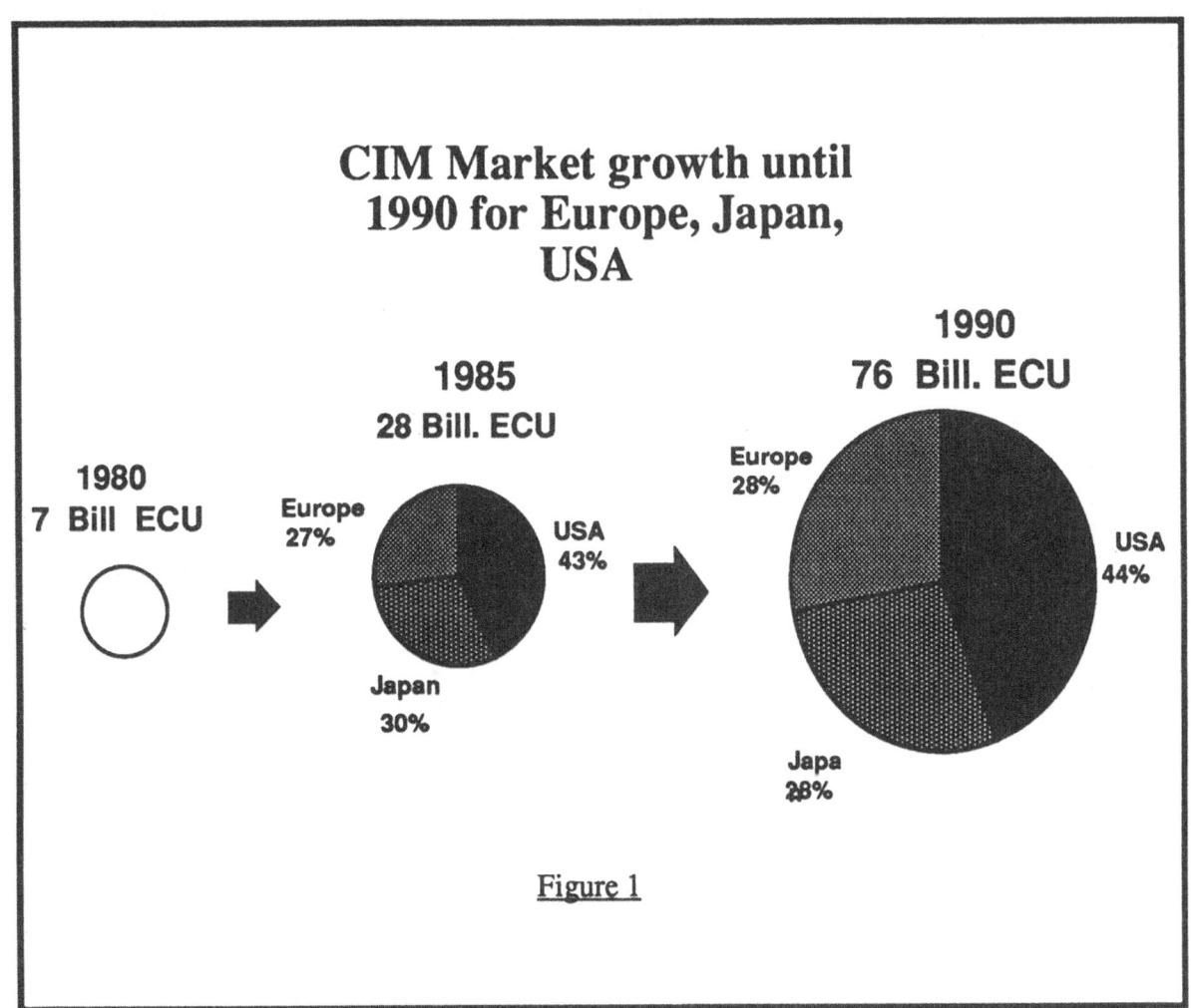

Figure 1

Products are more and more numerous, reliable and inexpensive. Thus was published in January 1990 by EMUG a "MAP 3.0 PRODUCT OVERVIEW" [5] in which MAP products and vendors were listed and classified. In this document some major world MAP applications were also listed. In spite of some I*LAN installations (with LAC or FACTOR in FRANCE) in medium or "little big" enterprises or units, it is quite obvious that, in most of cases, such realizations are still the domain of major enterprises especially in the area of automotive or aerospace production where they very often deal with pilot shoopflor or plant only.

Does this mean that small and medium enterprises are not concerned with communication architecture and integration problem ? Does this mean that CIM is only a major enterprise concern?

The answer is obviously no (in a medium and long term perspective). But the integration and automation process in small enterprises are quite different to that in the larger ones due to their differences in terms of available resources more than requirements. Therefore, it is useful to examine first the approach to automation in small and medium enterprises.

2. AUTOMATION IN SMALL AND MEDIUM ENTERPRISES.

In order to observe how small and medium enterprises deal with their automatisation, we can firstly examine investment policy in the field of industrial data processing or automatisation eventually compared with that of the biggest enterprises.

Figure 2 a and 2 b give the average number of main frames, mini-computers and micro-computers compared to number of employees in industrial plant in 1989.[6].

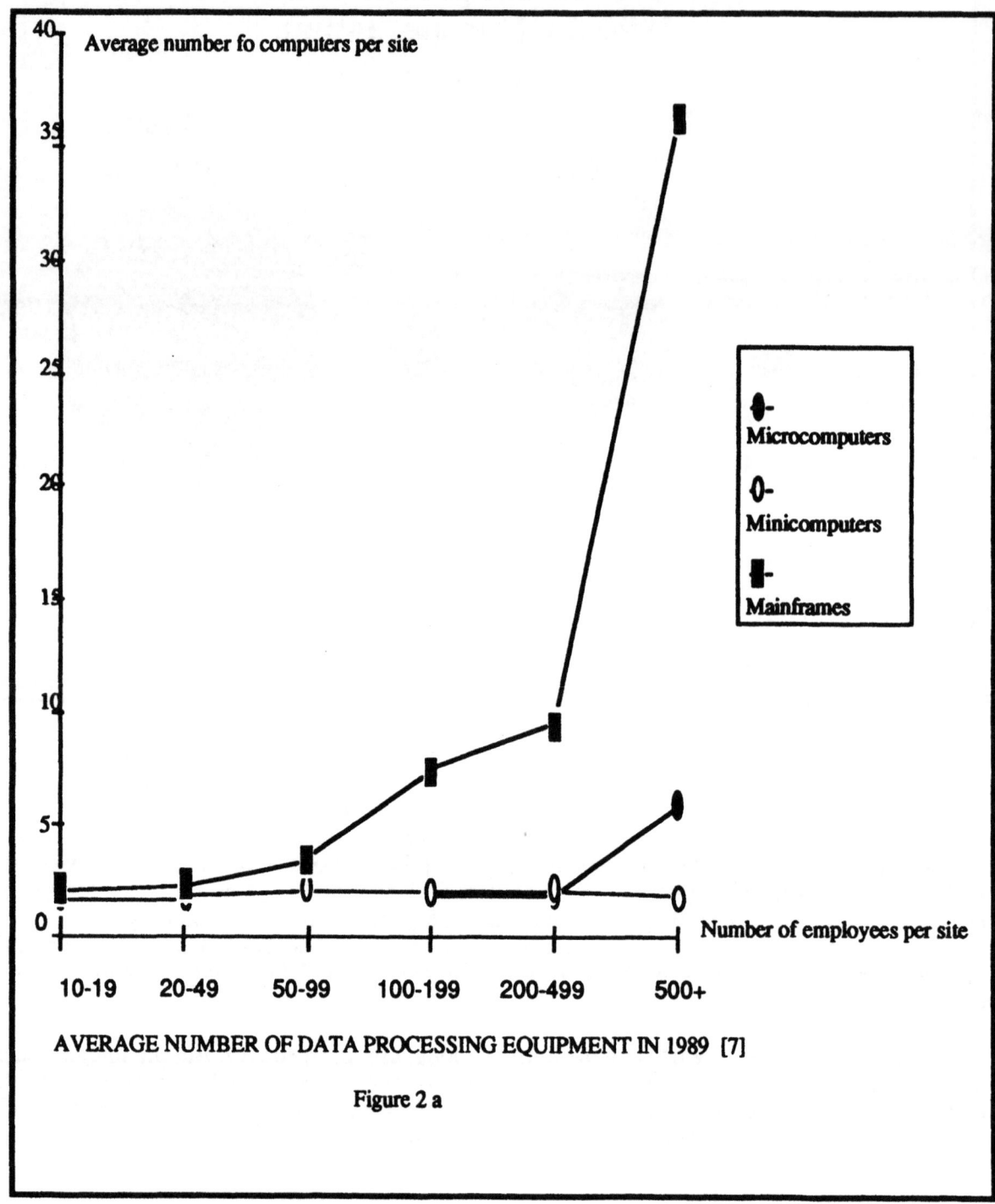

AVERAGE NUMBER OF DATA PROCESSING EQUIPMENT IN 1989 [7]

Figure 2 a

The Figure 2 b gives the number of computers in 1988 compared to this of 1989 for plant which have answered to the annual ADEPA poll in France.

	Total number		Average number per site	
	1988	1989	1988	1989
Micro-Computers	65 000	95 200	4;4	5,3
Mini-Computers	11 300	15 250	2,4	2,1*
Mainframes	6 400	9 500	1,8	1,4

Figure 2 b
Data processing equipment in industrial site [6]

Some conclusions can be gained from this data.The growth in term of computer is more than 35% between these two years. But the number of mini computers is (35%) not growing as rapidly of that of number of micro computers(46%) and even number of mainframe (48%).

In small and medium plants the number of minicomputers or mainframes appears as independant of number of employees (up to 500 employees per plant). This can be explained by four factors :

a)- In an industrial environment data processing equipment is related to the type of process and production line rather than the number of workers.

b)- The numbers provided in figure 2 b don't take into account the size or the price of equipment included in the class of minicomputers or mainframes, especially with respect to peripherical devices as terminals, printers,...

c)- Scale savings can be greater for these two categories of computers than for the personnal classe leading to the fact that the need of computer resources (in term of mainframe) doesn't increase in proportion with the size of the plant.

d)- Data processing resources in large enterprises are often centralized in their headquarters, while in small and medium enterprises the headquarters and shopfloor are often of the same site ; thus increasing the proportion of computer to personnel per site.

This clearly shows that small and medium enterprises are not lagging in term of data processing equipment. In addition with respect the number of employees, the chart clearly indicates that the ratio of micro-computers to personnal is greater in sites of less than 200 employees than in larger ones. In fact when productivity investments are in the affordable range (like micro computers) decision making in small and medium enterprises is very easy and fast They don't hesitate to improve their production be use of automatisation of computerisation.

Figure 3 gives some indication the distribution between very dynamic enterprises in term of automation or investments and others.[6].

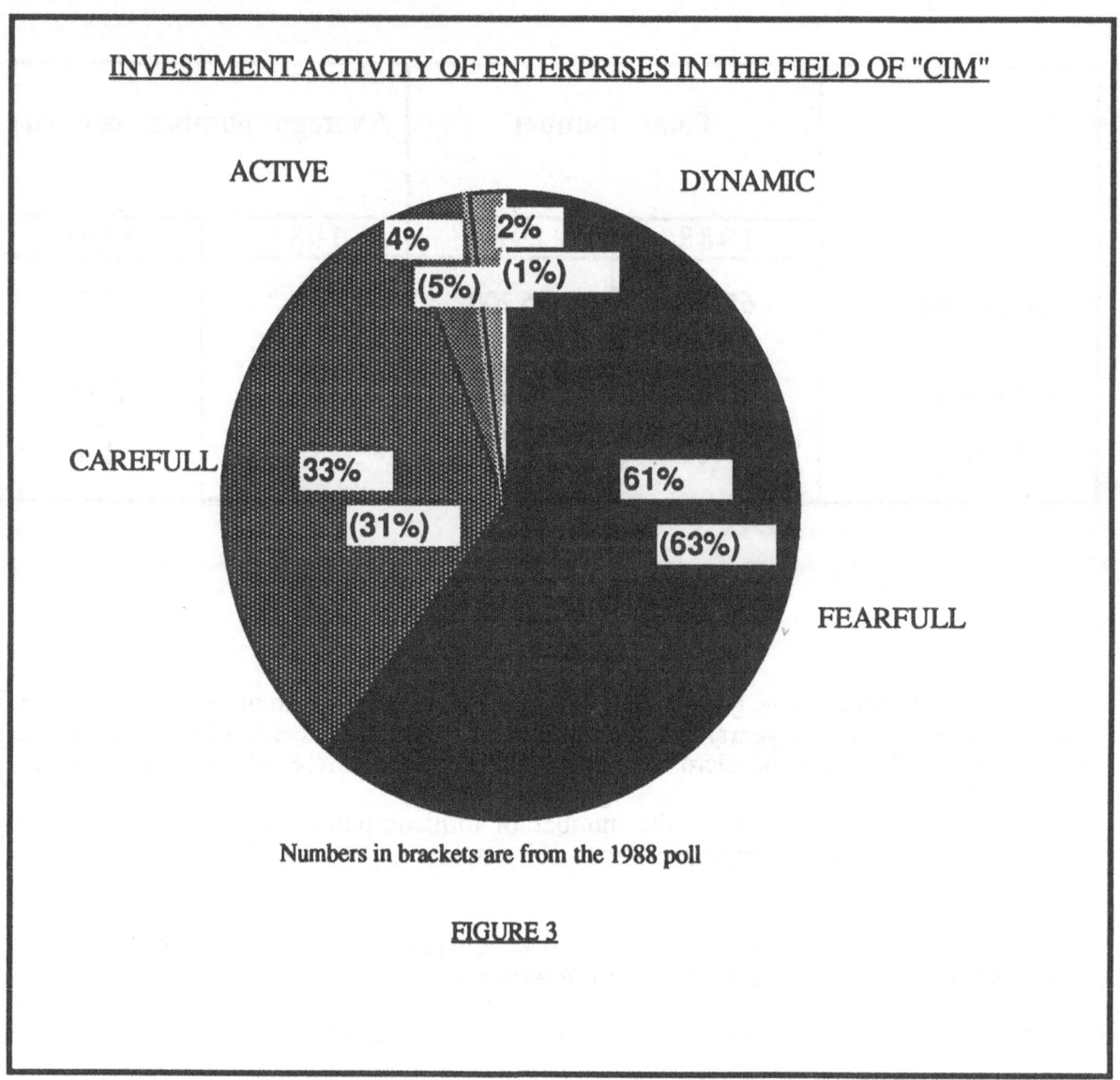

INVESTMENT ACTIVITY OF ENTERPRISES IN THE FIELD OF "CIM"

ACTIVE

DYNAMIC

4%

2%

(5%)

(1%)

CAREFULL

33%

61%

(31%)

(63%)

FEARFULL

Numbers in brackets are from the 1988 poll

FIGURE 3

Thus the small and medium enterprises have the same perspectives and concerns in term of industrial automation, productivity improvement, cost reduction as the larger ones, but they don't have the same capacity of investments. This fact will appears of major importance in their behaviour facing I*LAN's implantation and computer integrated manufacturing.

Figures 4 and 5 give the distribution of investments (in percentage, by asked enterprises) during the last twelve months and those foreseen by the same enterprises for the next twelve months, among for classes of applications for both hardware (Figure 4) and software (Figure 5) [6].

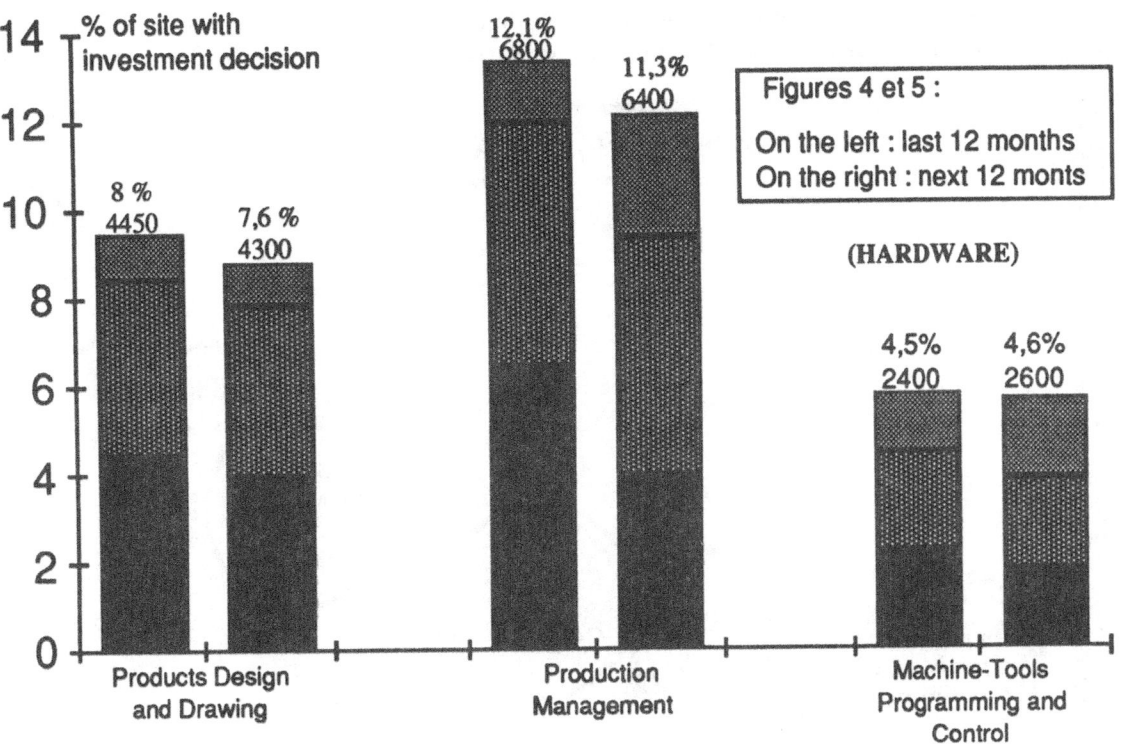

NUMBER (or %) OF SITES WITH INVESTMENT DECISION IN THE FIELD OF CIM

Figure 4 [6]

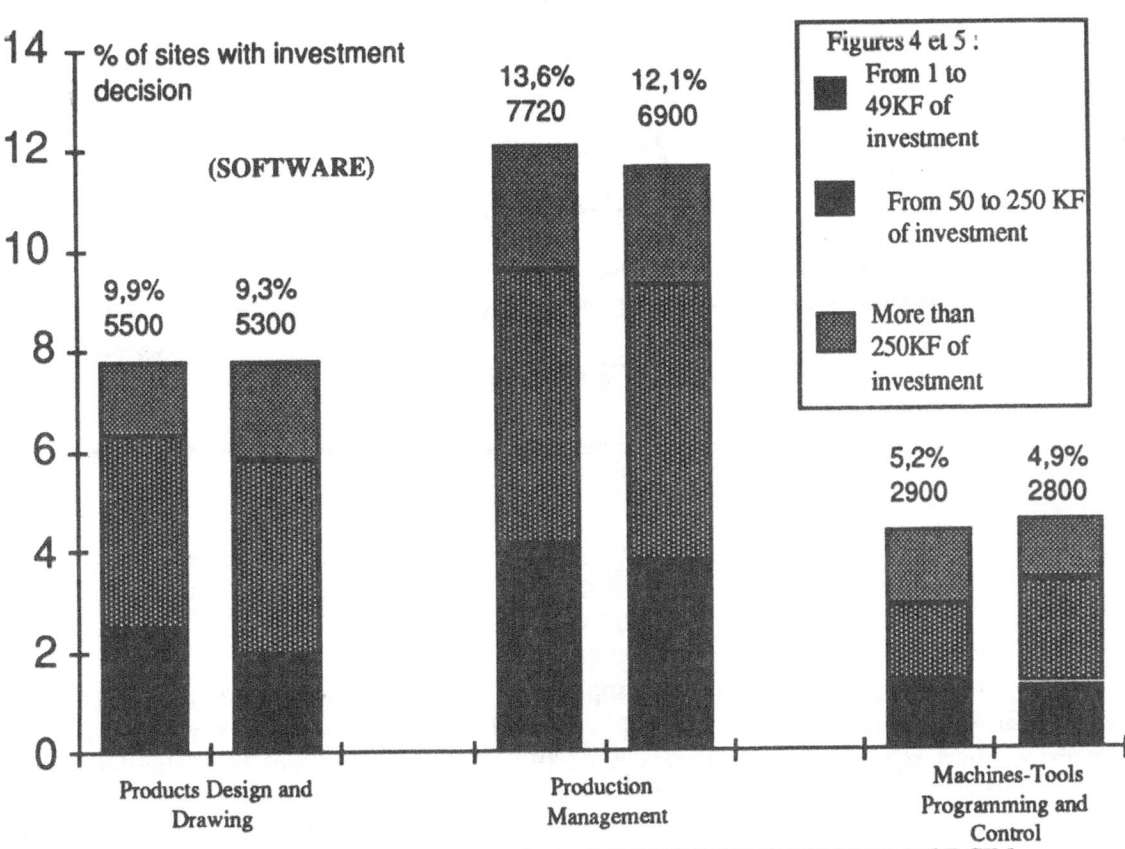

NUMBER (or%) OF SITE WITH INVESTMENT DECISION IN THE FIELD OF CIM

Figure 5 [6]

Further more Figure 6 shows how different techniques to be used in order to improve industrial productivity can be placed with respect their cost and their delay of efficiency.[7].

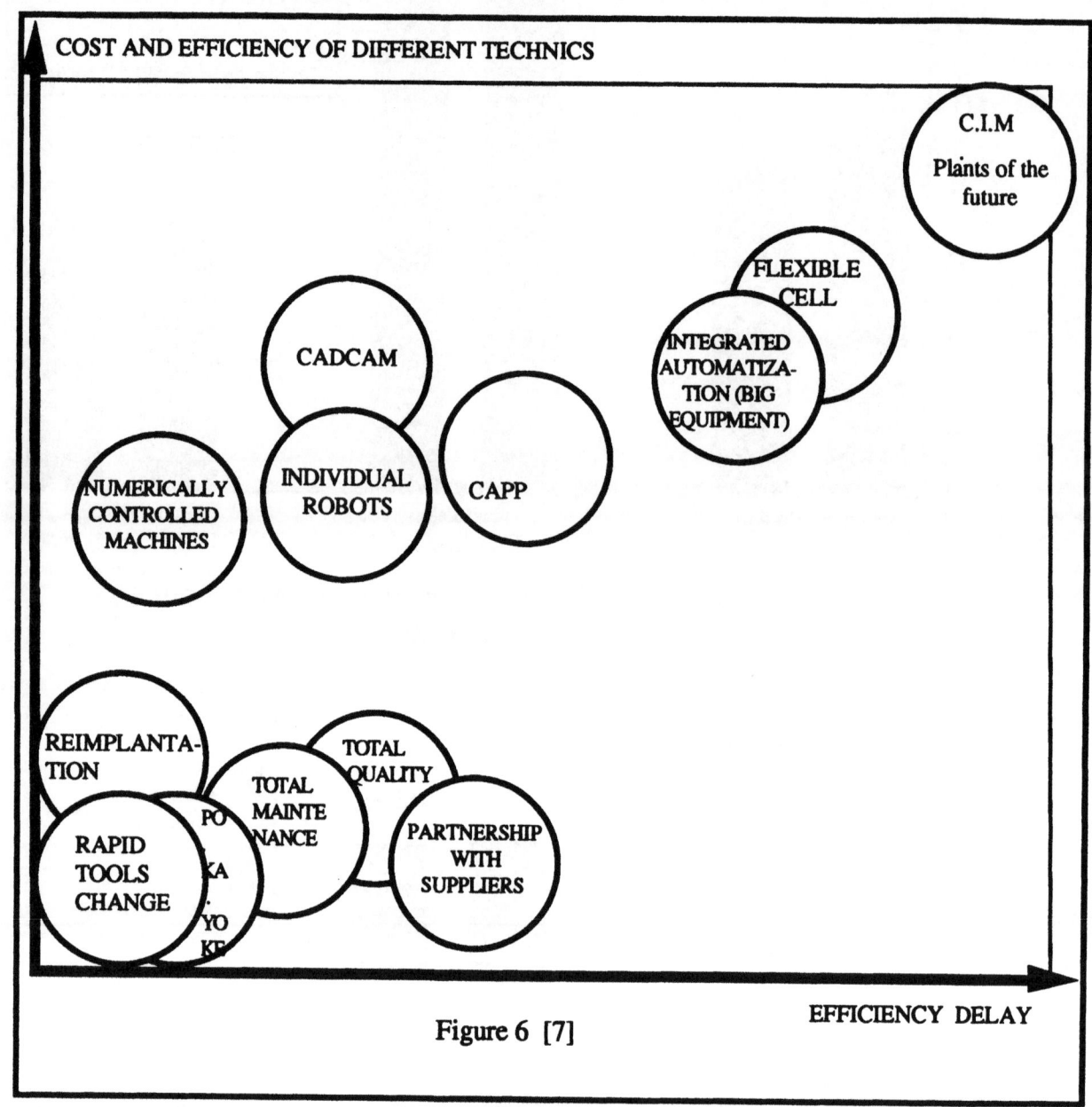

Figure 6 [7]

This picture clearly shows that some simple, low cost techniques with very quick efficiency like rearranging layout, rapid tools change are much more attractive for small and medium enterprises than expensive and complex ones, like I*LAN'S, flexible cells, integrated computer manufacturing. Very often indeed the results and efficiency of these complex techniques takes a long time to be obtained. Adjustments, tuning are very difficult and financial success is not always the result as can be seen on Figure 7 based on observations made by the CETIM's production engineering Department in charge of productivity improvement in mechanical manufacturing enterprises.

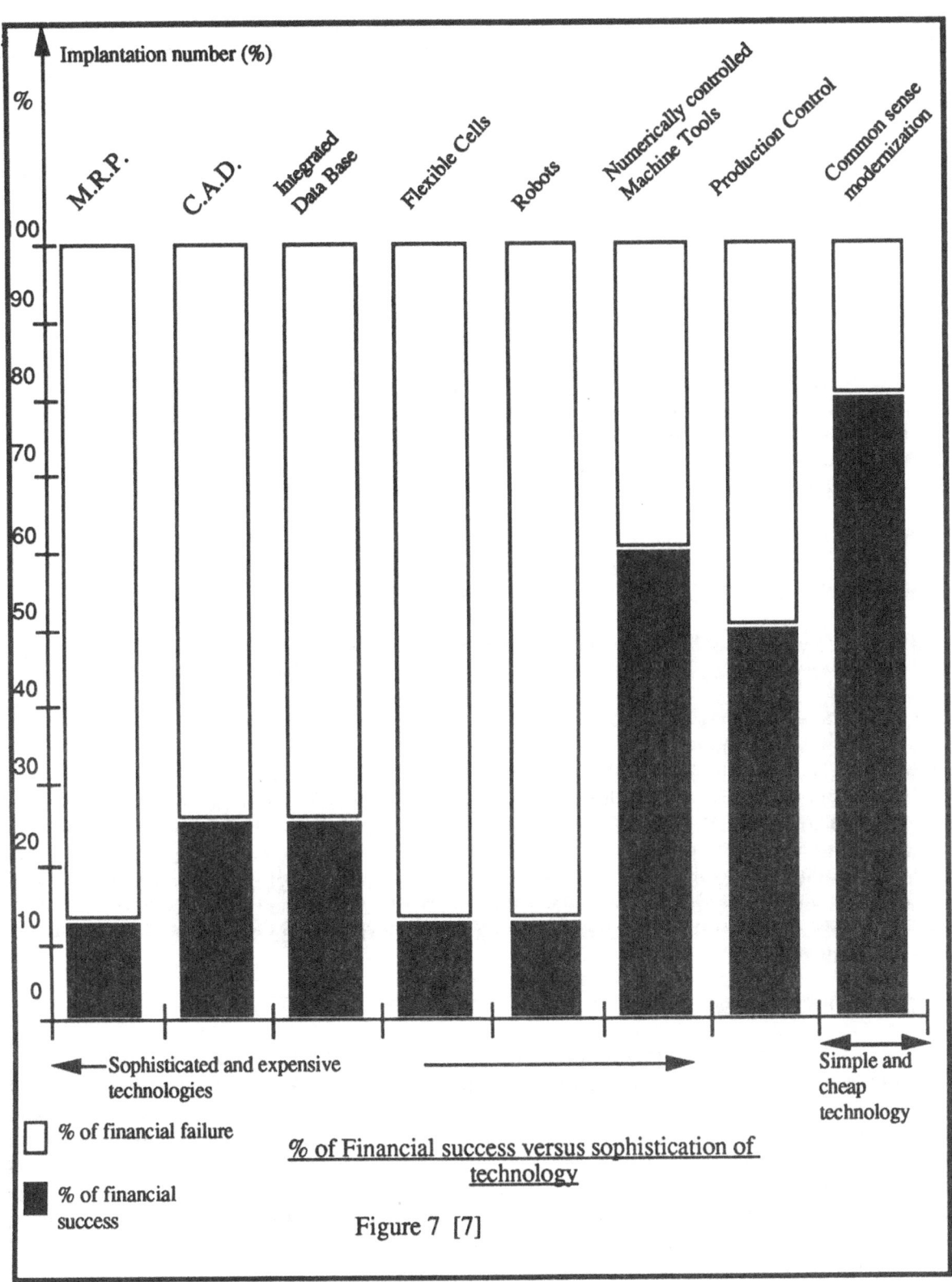

% of Financial success versus sophistication of technology

Figure 7 [7]

37

From those observations it clearly ,appears again that objectives, perspectives and concerns do not differ between small, medium and large enterprises. Nevertheless the approach of small and medium enterprises to communications architecture design, for I*LAN's implementation and for computer integrated manufacturing is always careful and seems to be sometimes fearful. This isn't due to a lack of perspective or innovation capabilities but to the specific constraints under which they are running faced to a very large, complex but sometimes incomplete and sketchy supply.

3. CONSTRAINTS OF SME IN AUTOMATION IMPROVEMENT.

For small and medium enterprises the constraints, and the outcome, of communication architecture building and computer integrated manufacturing are often specific and arise from two kinds of factors :

3.1. - Constraints due to the lack of resources.

It is obvious that, in spite of continuous falling cost of computer's hardware, some techniques, software and components, in the field of industrial automation are still very expensive. Even though substantial decreases are expected by some people, it is nevertheless clear that the cost of realising minimal application with this type of technology is still greater than the investment capabilities of small or medium enterprises. It can indeed be equal to several times their annual investment budget, and pay-back isn't so clearly established.

Another kind of resources is missing too and the lack of it is much more serious than asset deficiency. This is very often forgotten or neglected by people who are afraid of apparent SME deficiency in this field. This means a lack of human ressources..

All this technology is still very complex and rapidly evolving, new products on the market are more and more numerous. Nevertheless supply is very often incomplete or inadequate to the problems which are to be solved. Managers of SME are afraid of fast obsolescence of techniques and products proposed. They are also afraid to add new devices or equipment which are different from the other already present and thus increase again heterogeneity of their installation and increase too their communication missing. Facing this situation small and medium enterprises don't have enough human availability in term of either time and competence. These techniques are rather new and about their evolution. Then to make right choice in the every time, to foresee difficulties and traps, enterprises need people who is very well informed of those technologies, their evolution and development. That needs a strong training effort and is time consuming if competence should be maintained during the whole cycle life of the project well in small and medium enterprises more than in larger ones. But engineer time is the scarest ressource in small and medium enterprises.

3.2.- Other difficulties

In most cases, small and medium enterprises don't start their aumatisation process with the implantation of I*LAN's and communication systems. They always have (like bigger enterprises) existing computerized or automated applications which use isolated heterogeneous equipment. For exemple computers, NC machines, CAD stations, process controllers... Then to improve productivity, they want to connect harware and software to another in order to avoid manual operations between those two different applications. They require to do that with very simple and inexpensive solutions because they don't have ressources to launch in bigger investments and this simple connection can rapidly provide important savings and returns or solves very high difficulties in production process. When visiting small enterprises of mechanical sector, CETIM's engineers very often encounter these type of requirements or questions :

- How can I transmit files from finite element software runing on one specific hardware to a CAD system proceeded one another hardware and vice versa ?

- How can I connect rapidly, inexpensively and simply an actuator to a control cell or micro computer ?

The answer can sometimes be very simple at least for hardware or low layer software part : RS232, Kermit protocole can sometimes be components of this kind, fast, simple, inexpensive but very often incomplete solution. Specific software must be developed in these cases to ensure all the functions required by the application. This will be the source of most of the difficulties later on when software and hardware is maintained or obsolete. Difficulties will also happen when new applications must every time be connected to those already developed.

Larger enterprises sometimes also use this approach and lose money and time (in a long term view point) when maintening and evolving very old complex and inconsistant applications. Nevertheless they can also have a more comprehensive approach, it is possible to start with a global thought and design a whole CIM and communication architecture. They can test technology, concepts and architecture in pilot shoopfloor or plant, then they can make their choices after their test results. They can evaluate all kinds of consequences of such an implantation : economic, organizational and even human factors if they are aware of those.

Small and medium enterprises cannot proceed in this way, they don't have time to spend on very large or in depth preliminary studies and evaluations.They haven't series of shoopfloors or plants between which an adequate choice can be made to develop long term experimentation or studies, before important test site installation which will carry out substantial means in terms of investment (hardware and software) and human ressources. They don't have sufficient time and ressources for this approach.

Another method can be partial and experimental transformation of their production line in order to evaluate, in an actual situation, choices in term of architecture, hardware and software. This approach seems very interesting, but can be very expensive in the case of failure, not only due to lose of investment and time, but especially because of serious perturbations which can occur in their daily production, such as adjustment, breackdown, failure or anomaly.

Does that mean that no solution is available for small and medium enterprise who wish to undertake an integration of their different automated systems, with an whole view point of obtaining ultimately a real computer integrated manufacturing system ? Do these enterprises wait until thinks are fully stabilized ?

The answer is probably "NO". Thus CETIM is putting down a methodology to design and to implement I*LAN's, communication architecture and to ultimately integrate all automatised equipment (or non automatised) into a whole coherent integrated manufacturing system adapted to SME requirements and to a rate of investment and advancement that stay compatible with resources. CETIM will also install a National Pilot Shoopfloor within this methodology which will be tested and evaluate in actually production situation.

This shoopfloor already includes one milling center (Huron Graffenstaden), two turning centers, presetting tools and stations and CNC machine tools. Most of these tools are directly connected (or through sub'LANS like UNITELWAY) to a MAP broadland I*LAN which will be in the nearfuture extended to CAD, Method and Production Control Department. Ultimately the shoopfloor MAP system which is subject of cooperation contract betwen CETIM and the BULL Company will be connected to the Ethernet main LAN of the CETIM.

The whole project is obviously not yet achieved and a great part of it is still under installation or development but most of concerned equipments on the shoopfloor is already connected and some applications like part programmes or tool gauge transmission between milling center and presetting tools will be soon available. On the other hand CETIM has installed in the Saint-Etienne site a demonstration flexible cell which is already used for a few years for training courses. Then some principles can be already suggested below, provided from such a methodology.

4. HOW TO START CIM SYSTEM WHEN ENTERPRISE IS SMALL.

The main rule in this approach is the following :

Start very modestly with a very simple application from which the results will appear rapidly and will be obvious . The application must be designed and implemented within a whole perspective, with options and choices which must be both anticipative (from an technical and business viewpoint) and perennial ones.That could also be the only rule of the methodology. From this all the other principles, purposes and suggestions which are listed below can be deducted.

Make anticipative and perennial choice doesn't hide a contradiction. That means that technical options must not be too advanced, nor based on unstabilized products and very fast evolving techniques. That also means, on the other hand, technical option must not perpetuate old and obsolete solutions which can provide more and more difficulties . That doesn't exclude, from the new application, existing equipment and even specific developments. But this type of decision must be made with full knowledge of the situation. That will be an interim solution which will prepare for further developments and investments according to the whole perspective and architecture choosen for the enterprise and to the products or technics evolution already expected from the market.

With this kind of approach, new applications, new investments can be made progressively, step by step. Each application must be designed, realized and implemented one after the other. It must be made according to the rules and principles previously described.It must be compatible with the choices and architecture defined at the beginning. Nevertheless in these new applications existing equipment can be used, very simple and limited specific developments can be considered if they are strongly justified, positioned in a long term perspective, if they don't counteract the essential choices or options and if they, ultimately, are expected to be temporary ones.

All these studies and thoughts are very long, complex and sometimes a little hazardous. Small and medium enterprises don't always have, as previously explained, enough human ressources within their organization to lead extensively so long and complex investigations and studies. Fortunately two major elements can be helpful for small and medium enterprises in their difficulties. The first one is the progress of international standardization works in the field of CIM and advanced manufacturing technologies. The second is the help of industrial support center as applied research or technical centers and engineering companies.

4.1 - Help from International and European standardization.

Very big activity is currently provided by international, european and national organization in the field of computer integrated manufacturing and advanced manufacturing technologies (AMT). Works have been done on ISO TC 184 Industrial automation and integration system, in IEC 65 and in CEN/CENELEC especially through its ITAEGM (Information Technology Advisory Expert Group on Manufacturing) common group with ETSI.

Other organizations like MAP/TOP world federation, EMUG, ... are also working on those subjects and very often their members are also participating to official standardization commission. EEC started many years ago a very large research program ESPRIT in which a lot of projects are connected to prestandardization works. Those very often lead to standardization proposals. Concurrently considerable efforts are provided in the domain of certification for these, type of products and techniques. This is of major interest for users and especially for small and medium enterprises. Firstly, standardization will contribute to prices decreasing due to the worldwilde market extension. Secondly, systematic choices of standardized options or perspectives (if available) is the only way to get a reasonable guarantee that investments will have a good lifetime. But standards are more and more complex. They are very hard and difficult to read for non specialist people ; they provide an high number of options which can be in fact sometimes incompatible with one another (some standardisation are working because of that on profiles which can provide some guarantee of "interoperability"). For these reasons small and medium enterprises will need another kind of help described below.

4.2 - Help through outside services.

4.2.1 - Industrial support centers

The better way to perform this comprehensive thought and design, which is founded on very strong theoritical and practical knowledge of those very specific techniques (which aren't their business) is for small and medium enterprises to call services of industrial support organizations. That can be first an industrial applied research centers which very often are also in charge of technical knowledge transfert and support to industrial environment. This type of industrial applied research centers set out pilot shoopfloors with the purposes of characterising and demonstrating those techniques in an actual production environment. They can thus evaluate implementation difficulties, products and techniques availability and stability, optimal conditions of utilisation. The small and medium enterprises will have then access as the larger ones to pilot shoopfloor, they cannot hope to build or purchase on their own.

4.2.2 - Engineering and services companies

Small and medium enterprises can also call on the services of engineering enterprises which are more and more called for this type of design and installation by SME in various sectors. Thes engineering enterprises get then a very good practical knowledge of those problems, they are also more and more involved in the follow-up of advanced manufacturing technologies and their standardization. Then, like the technical centers mentioned above, they can greatly help small and medium enterprises in the comprehensive thought of computer integrated manufacturing. In co-operation with the enterprises they can perform process and business models, simulation and computer integrated manufacturing architecture design according to whole and long term purpose and perspectives of the enterprises. After this, they must be in charge of the follow-up of stepwise realization, overtime simple application after simple application to ensure that each new implementation is not in opposition with previous choices and to help enterprises to manage difficulties due to the complexity of those techniques. That means especially existing equipment, the possibly lack of any product, and so on.

5. CONCLUSION

In conclusion small and medium enterprise aren't too lagging in the modernisation process as sometimes heard. Facing the computer integrated manufacturing and communication architecture, they get the same objectives, problems and concerns as the major ones . But due to their size and their lack of resources they must have a progressive approach and call help of technical centre (and their pilot shoopfloor) or engineering enterprises. Then they put their step by step realizations into a whole perspective and for a long term viewpoint based on a prerequisite thought and design of the future computer integrated manufacturing architecture of enterprise.

6. ACKNOWLEDGEMENT.

I would especially like to thank M. MORONVAL and P. PADILLA who were helpful to me by giving their advice, remarks based on a high comprehensive and practical knowledge of small and medium enterprises in the mechanical engineering industry, through a very high scientific background in the field of mechanical design and production.

7. REFERENCES

[1] MAP : Manufacturing Automation Protocol, Version 3.0 implementation release; April 7, 1987

[2] CEN/CENELEC/CEPT Information Technology Steering Committee (ISTC) Memorandum M-IT-04

[3] CEN/CENELEC ENV 40003 to be published (April 18, 1990)

[4] EMUG MAP 3.0 product overview January 25, 1990

[5] Schwimann, H., "Der weltweite stand von CIM", Technische Rundschau, Vol. 38, pp. 122-128 (1986)

[6] ADEPA : "Enquête annuelle sur les équipement en productique" ADEPA - rue Périer - Montrouge - France.

[7] Padilla P., "Amélioration de la productivité des centre d'usinage" CETIM Informations, n°108

INTERNATIONAL INITIATIVES

CURRENT STATUS AND OUTLOOK FOR FAIS

September , 1990
International Robotics and Factory Automation Center
(IROFA)

1. INTRODUCTION

1.1 WHAT IS FAIS?

FAIS stands for "Factory Automation Interconnection System," a research and development project being pursued by the International Robotics and Factory Automation Center (IROFA) under a commission from the Ministry of International Trade and Industry. The purpose of this project was to develop unified interconnection technology for the automatic equipment used in manufacturing.

A budget of ¥300 million was set aside for the FAIS research and development project. The FAIS development team was located within IROFA, and the project was pursued jointly by industry, academia and government.

The yearly schedule for the project is outlined below:

1.   FY87 (Nov. 1987 — Mar. 1988)
        Analysis of FAIS survey; conceptual design.
2.   FY88 (Apr. 1988 — Mar. 1989)
        Development of FAIS Cell-Net Implementation
        Specifications; design of verification and evaluation
        system.
3.   FY89 (Apr. 1989 — Mar. 1990)
        Tests using the verification and evaluation system to
        verify the interconnectivity of FAIS.

1.2 BACKGROUND TO THE FAIS PROJECT

Studies prior to the inception of this project, as well as the survey and analysis performed during the first year, yielded the following information regarding the status of manufacturing system networking:

1. Manufacturing systems are moving away from the automation of single processes and more in the direction linking up and integrating with the information management systems of the entire company. To do this it will be absolutely necessary to interconnect a variety of equipment from a variety of vendors. MAP (Manufacturing Automation Protocol) has been put forward as the specifications for doing so.
2. MAP commercialization is starting with "Full-MAP" systems, which are positioned as the trunk network. Mini-MAP systems, which are suited to networking in manufacturing facilities are still at the stage of technical discussion.
3. Japanese corporations tend to take a bottom-up approach to their work, so the demand is strong for a factory-based multivendor network that can operate in real time.

1.3 SCOPE OF RESEARCH AND DEVELOPMENT

Given the situation described above, it was decided that the FAIS project would concentrate on the research and development of technology to integrate and provide easy interconnectivity between equipment from different vendors (Fig. 1) at the cell-level. It was decided to use optical fiber cables for data transmission in addition to coaxial cables as Japanese users are more interested in the former.

The cell-level networks that were the focus of the project are known as "FAIS Cell Networks" and their detailed communications protocol is known as the "FAIS Cell-Net Implementation Specifications."

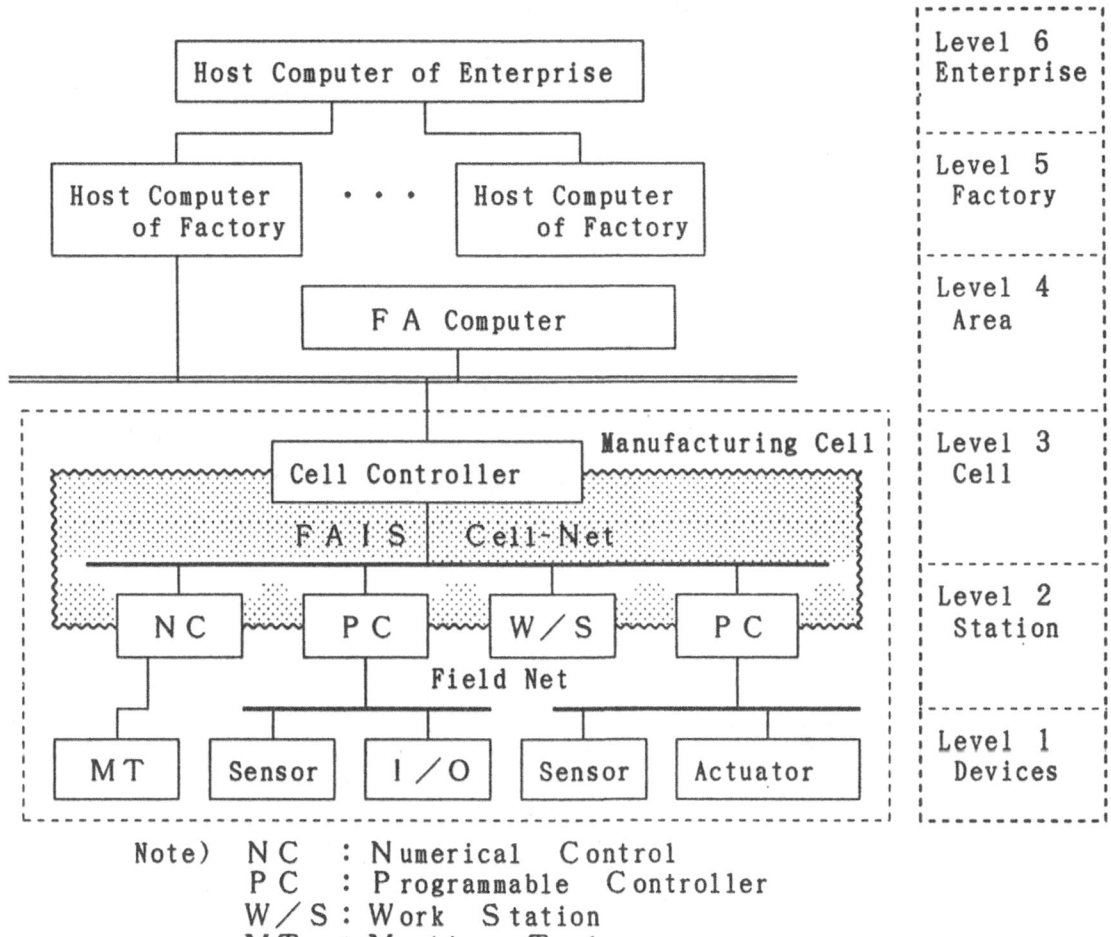

Note) N C : Numerical Control
      P C : Programmable Controller
      W∕S : Work Station
      M T : Machine Tool

Fig. 1 Subject of F A I S Research & Development

2. FAIS CELL-NET IMPLEMENTATION SPECIFICATIONS

The second draft of the FAIS Cell-Net Implementation Specifications was published in March 1989. This was followed by simulation tests and the subsequent evaluations, which resulted in the publication of Version 1.0 in March 1990.

2.1 THE FAIS ARCHITECTURE

FAIS adopts a three-layer architecture (Fig. 2) similar to the Mini-MAP system. This was done to ensure compatibility with MAP. In the FAIS Cell-Net Implementation Specifications, the Physical and Data Link layers (LLC sub-layer and MAC sub-layer) are located in the lower layers, and the Application layer in the upper layer.

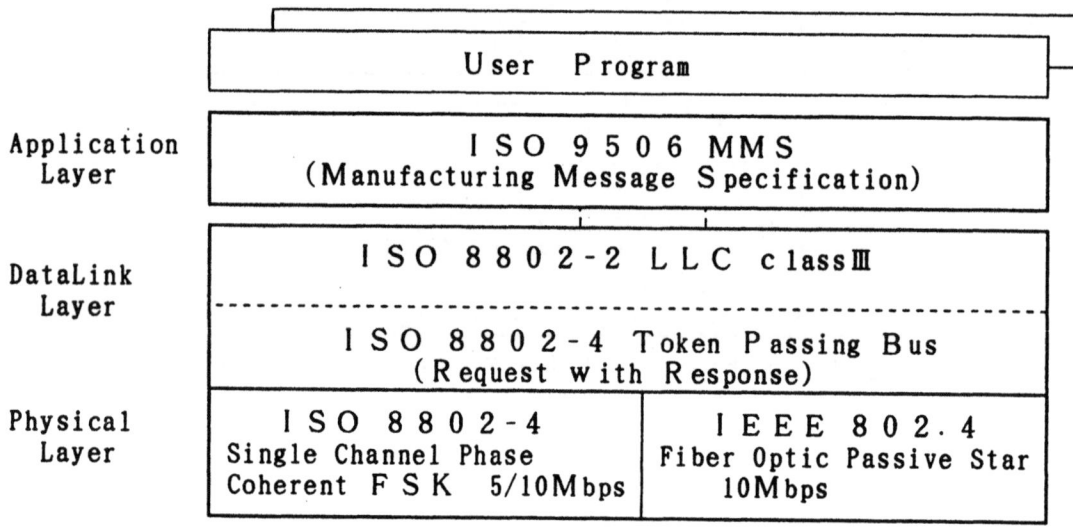

Fig. 2   Architecture of F A I S

In creating the FAIS Cell-Net Implementation Specifications, the functions specified for Mini-MAP and MMS were subsetted under the following guidelines:

1. Only token-holding-station is supported as a node type.
2. Both association-oriented and associationless communication modes are supported.
3. Only point-to-point communication is supported.
4. MMS services are within the range of the MAP 2 Implementation Class defined in MAP 3.0- Implementation Release Subject to Errata Changes.

2.2 UPPER LEVEL IMPLEMENTATION SPECIFICATIONS

(1) Scope of Upper Layer Specifications

The specifications cover MMS providers, and detailed specifications were formulated for MMS protocol and services. The specifications do not cover MMS users (servers/clients). Detailed specifications were created for the following eleven MMS services:

- Initiate
- Conclude
- Abort
- Cancel
- Reject
- Status
- GetNameList
- Identify
- Status
- Write
- GetVariableAccessAtrributes

(2) Logical Model of the Upper Layer

The FAIS upper layer was specified based on the logical model shown in Fig. 3. This model is based on the definitions of PCA (Process Communication Architecture), and comprises MMPM, ACM and APM auxiliary protocol machines.

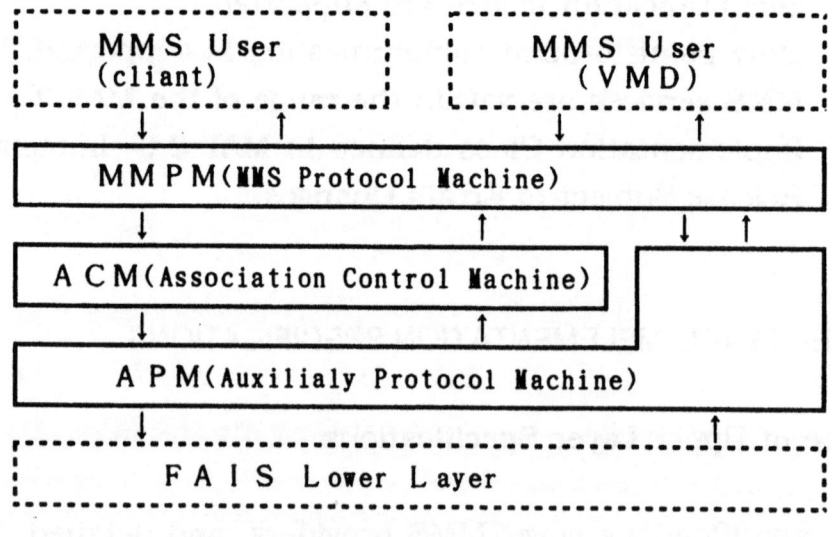

Fig. 3　Upper Layer Logical Model of FAIS

(3)　Communication Modes for Upper Layer Protocol

LSAP-orientation and PDU-orientation are specified in order to distinguish between association-oriented and associationless communication modes.

There are three types of LSAP-orientation: association, associationless and association unspecified.　The type to be used must be specified.

There are six types of PDU orientation, which are listed below. They verify the communication mode between the sending and receiving nodes and control the application association.

- ASOC_ORIENTED_PDU
- INITIATE_PDU
- ABORT_PDU
- ASSOCIATIONLESS_PDU
- CONCLUDE_PDU
- REJECT_PDU

(4) PDU Orientation in Upper Layer

PDUs in the upper layer (FAISmmsPDU) contain auxiliary protocols in front of the MMS pdu to indicate their orientation and specify the context of abstract syntax.

(5) Detailed Specifications for the Upper Layer

Detailed specifications have been made for the implementation of communications functions, the MMS parameter subsets which must be implemented, and abstract syntax and parameter values. The major specifications are listed below:

1. Both association-oriented and associationless communication modes are required.
2. Both association-oriented and associationless LSAP-orientations are required.
3. The maximum length of upper layer PDUs (FAISmmsPDU) is 1024 octets.
4. Only the fixed length forms specified in ISO/DIS 8825.2 are used encoding rules.

2.3 LOWER LAYER IMPLEMENTATION SPECIFICATIONS

(1) LLC Sub-layer Specifications

The following detailed specifications were made concerning the implementation of ISO 8802-2 LLC Class III.

1. The implementation is required to be able to receive Type 1 commands and send Type 1 responses.
2. The following points regarding the interface with the upper layer were also stated clearly:

- The LSAP value used depends on the upper layer application.
- The resynchronization of LLC Type 3 is caused by the upper layer.
- The types of status information to be supplied to the upper layer were also noted.

(2)   MAC Sub-layer and Physical Layer Specifications

The following detailed specifications were made regarding the implementation of the ISO 8802-4 token passing bus method.

1. Service classes and accesses classes linked to the priority of communications are required.
2. Request_with_Response functions are required. Assumed in this implementation are the specifications for data transfer and data exchange with confirmation functions for LLC Type 3.

Parameter values for the MAC sub-layer depend on the size of the system. Sample settings are included for an assumed system of transmission line length 500m — 1km, with about 32 nodes.

For the physical layer it is recommended that DTE-DCE interfaces, loopback functions and initialization sequences be implemented.

(3)   Fiber Optic Specifications

Two systems, high-sensitivity and moderate-sensitivity, are assumed for the model system (Table 1). Values specified conform to IEEE 802.4.

Table. 1 Model System for Fiber Optic Implementation Specification

	Items	Moderate Sensitivity	High Sensitivity
①	Topology	Passive Star	Passive Star
②	Number of Port	8	3 2
③	Maximum Distance between Node & StarCoupler	5 0 0 meter	5 0 0 meter
④	Maximum Number of Splices between Node & StarCoupler	2	2

Specifications at the physical layer also cover the fiber input level, receiver power range, allowable transmission loss and optical silence level for optical fibers of $50/125\mu m$ (core diameter / clad diameter), the type most commonly used in Japan.

For media, specifications were made for test fibers, fibers used, connectors, splicers, star couplers and reflected light levels.

3. SIMULATION TESTING

Tests were made to verify that systems developed by various companies in accordance with FAIS Cell-Net Implementation Specifications could actually communicate with each other. The results verify the content of FAIS Cell-Net Implementation Specifications and confirm interconnectivity.

3.1 TEST PROCEDURES AND PARTICIPATING COMPANIES

The thirteen companies listed below submitted a total of twelve systems for the simulation tests. The systems were connected to carrier band and optical (moderate- or high-sensitivity) transmission lines.

Participating Companies*

Fanuc Ltd.

Fuji Electric Co., Ltd.

Fujitsu Ltd.

Hitachi, Ltd.

Hitachi Cable, Ltd.

Matshushita Electric Industrial Co., Ltd.

Mitsubishi Electric Corp.

NEC Corp.

Omron Corp.

Sumitomo Electric Industries Ltd.

Toshiba Corp.

Yamatake-Honeywell Co., Ltd.

Yokogawa Electric Corp.

*(Alphabetical Order.)

In order to verify that all systems had correctly met the FAIS Cell-Net Implementation Specifications, the first test checked communications between systems for each of the service primitives specified for the upper and lower layers. Following this, interconnectivity between applications was tested using communications functions already confirmed to be correct. This allowed the interconnectivity of the FAIS Cell-Net Implementation Specifications to be verified and evaluated. The first battery of tests are referred to as "implementation tests," the second "demonstration tests."

The implementation tests were first run between only four nodes to minimize the effects of mistakes in the specifications and differences in their interpretation. This not only allowed us to discover mistakes in the specifications, it also enabled us to establish reference nodes for which correct implementation had been confirmed. These 4 nodes formed the basis for the 12-node individual tests which followed.

3.2 LOWER LAYER IMPLEMENTATION TESTS

The basic connectivity of all systems participating in the tests was confirmed as regards the implementation of the LLC and MAC sub-layers.

(1) Description of Tests

Tests for the MAC sub-layer and below were designed to confirm: 1) that the initialization of the logical ring, entrance (induction) and separation are all performed correctly, and 2) that the logical ring operates in a stable manner.

Tests for the LLC sub-layer were designed to confirm: 1) that the data transmission service to the upper layer operates correctly, and 2) that LLC sub-layers can be connected to each other.

(2) Test Methodology

Tests for the MAC sub-layer and below were performed by starting up the MAC sub-layer directly from a test program and observing on a LAN monitor the actions of the MAC sub-layer and the physical layer.

Tests on the LLC sub-layer were performed with a test program that initiated transfer of Type 1 and Type 3 data. Observations were made with a LAN monitor hooked up to the transmission line and the receiving node.

(3) Tests of the Fiber Optic System and Physical Layer

Tests for the optical modem were designed to confirm: 1) the encoding of transmission data and its modulation into optical signals, and 2) the demodulation from optical signals and decoding of the data. Tests for the optical transmission line were designed to check loss from optical fibers and star couplers.

The Optical Transmission System Applicability Test Methodology discussed in IEEE802.4J4 was referred to in the optical modem tests.

3.3 UPPER LAYER IMPLEMENTATION TESTS

(1) Description of Tests

The forty-four tests listed below were performed in both the association-oriented and associationless communications environments for all MMS services for which there are detailed FAIS specifications. This was done in order to check the PDU transmission functions and coding and decoding functions of MMS providers.

Table. 2   Test Items for Upper Layer Test

Communication Envirnment	MMS Services	Number
Initial Condition	Initiate Conclude Abort	3 2 2
Association-Oriented	GetNameList GetVariableAccessAttributes Read Write Status Identify Cancel	3 3 4 4 2 1 2
Association-less	Read Write Status Identify Cancel	6 5 2 2 3

(2) Test Methodology

Upper layer tests were performed by having a test program issue MMS services. PDU encoding was confirmed using a LAN monitor connected to the transmission line, and decoding was confirmed from the output of the receiving primitive on the receiving side. The test program was created to test both the client roles and the server roles of the MMS providers.

(3) Results

The twelve systems supplied by the thirteen companies were connected together as shown in Table 3. It was confirmed that the systems conformed correctly to all FAIS Cell-Net Implementation Specifications.

Table. 3   Test Results of Upper Layer Tests

Client Role \ Server Role	Omron	Sumitomo Elec.	Toshiba	NEC	Hitachi Group	Fanuc	Fujitsu	Fuji Electric	Matsushita Elec.	Mitsubishi Elec.	Yamatake H.W.	Yokogawa Elec.
Omron	\	O		O			O		O			O
Sumitomo Elec.	O	\				O					O	
Toshiba			\				O	O		O		
NEC	O			\		O	O				O	O
Hitachi Group					\		O	O		O		
Fanuc		O		O		\			O			
Fujitsu	O		O	O	O		\					O
Fuji Electric			O		O			\				O
Matsushita Elec.	O					O			\		O	
Mitsubishi Elec.			O		O					\		O
Yamatake H.W.		O		O					O		\	
Yokogawa Elec.	O			O			O	O		O		\

Note)   O : Executed   Space : Not Executed

3.4   DESCRIPTION OF DEMONSTRATION TEST

Individual tests are designed to confirm the correctness of specific protocol. But just because different pieces of equipment pass them it does not necessarily follow that they can be connected together in actual applications. Communications between application programs simulating real-life applications were therefore used to generate imaginary situations and check interconnectivity on the application level. The following imaginary situations were used in the general tests.

(a)   <u>Control System for NC and Robot Equipment</u>
The cell controller controls a robot which mounts and removes parts and an NC (numerically controlled) unit which processes them.

(b)   <u>Control System for Electric Furnace and Handler</u>
The cell controller controls an electric furnace which heat-processes materials and a handler which feeds and removes the materials.

(c)   <u>PCB Mounting System</u>
The cell controller controls two different component mounting units which mount either ICs or discrete components on printed circuit boards.

(d)   <u>Water Management System</u>
The cell controller uses motors and valves to control the supply and removal of water to processing and cleansing facilities.

Six systems each were connected to moderate-sensitivity and high-sensitivity optical transmission lines and two situations each were enacted.

4. OUTLOOK FOR THE FUTURE

4.1 COMMERCIALIZATION OF FAIS

The FAIS project was wound up by opening the demonstration tests to the public for three days, from March 14 through 16, 1990. A survey taken of the visitors (about 1000 people) showed that approximately 30% of the respondents would like to see equipment which supports FAIS commercialized within the next two to three years. (See Table 4 for a list of products for which demand was greatest.) The ability which most of these products have to be interconnected should be able to be realized within the range of functions defined in the FAIS Cell-Net Implementation Specifications (Version 1.0).

The companies participating in the FAIS project are now working on the actual commercialization of FAIS products.

Table. 4 Demands for Commercialization of F A I S (multiple answer)

	Equipment Type					Total
	P L C	N C	Robot	FAComputer	Others	
Numb	1 4 8	7 3	8 6	1 8 3	2 3	5 1 3
%	2 9	1 4	1 7	3 6	4	1 0 0

4.2 WHAT IS LEFT TO BE DONE

Though the interconnection of multivendor equipment in the factory is now within sight, problems of time and resources mean that there are still a number of issues left to be tackled.

1. Expansion and Systematization of Specifications
   Version 1.0 deals, it could be said, with smaller systems on the practical level. In order to deal with the wider variety of fields to which they will be applied in the future, it will be necessary to

expand the scope of MMS specifications and add specifications for network management and object dictionaries.

2. <u>Evaluation Using Actual Equipment</u>
   Workstations and personal computers were used in the simulation tests to evaluate interconnectivity under the FAIS Cell-Net Implementation Specifications, but it is still necessary use real automated equipment to evaluate the practicality of applying FAIS to real manufacturing systems.

3. <u>Conformance Tests</u>
   As more and more equipment is developed based on the FAIS Cell-Net Implementation Specifications, conformance tests will be an effective way of preventing differences in specification interpretation.

4.3   THE FNE PROJECT

The FNE (FAIS Networking Event) Project got under way this year, charged with resolving the issues left over from the FAIS Project and promoting the results obtained.

Twenty-seven companies have expressed their desire to participate in the FNE Project at the present time.

Companies Participating in FNE*

    Fanuc Ltd.
    Fuji Electric Co., Ltd.
    Fujitsu Ltd.
    Furukawa Electric Co., Ltd.
    Hitachi Cable, Ltd.
    Hitachi, Ltd.
    IBM Japan, Ltd.
    JGC Corp.
    Kawasaki Heavy Industries, Ltd.

Matsushita Electric Industrial Co., Ltd.

Mitsubishi Electric Corp.NEC Corp.

Nihon Unysis, Ltd.

Nippon Steel Corp.

Okuma Machinery Works, Ltd.

Oki Electric Industry Co., Ltd.

Omron Corp.

Sharp Corp.

Shimizu Corp.

Sumitomo Electric Industries, Ltd.

Terasaki Electric Industries, Ltd.

Toshiba Corp.

Toyoda Machine Works, Ltd.

Toyo Engineering Corp.

Yamatake-Honeywell Co., Ltd.

Yaskawa Electric Mfg. Co., Ltd.

Yokogawa Electric Corp.

*(Alphabetical Order)

The main activities of the FNE Project are described below:

1. <u>Expansion of the FAIS Cell-Net Implementation Specifications</u>
   - Detailed specifications for MAP V3.0 network management.
   - Detailed specifications for MAP V3.0 object dictionaries.
   - Expansion of the scope of detailed specifications for MMS services.

2. <u>Development of Conformance Test Tools</u>
   Developed mainly for the specifications created for the FAIS and FNE Projects.

3. <u>Demonstrations Modeled on Actual Plants (FNE '92)</u>
   The goal of the FNE project is to check ease-of-use and interconnectivity in a form close to actual systems. The target for this is the beginning of 1992.

The FNE '92 demonstration will be performed for both factory-wide and cell-level applications as listed below:

- Factory-Wide: Production control, electric power control, heat-source monitoring and other utilities management.
- Cell-Level: Automated factory for laser processing of plastic sheets.
                      Process factory manufacturing products by annealing compound materials.

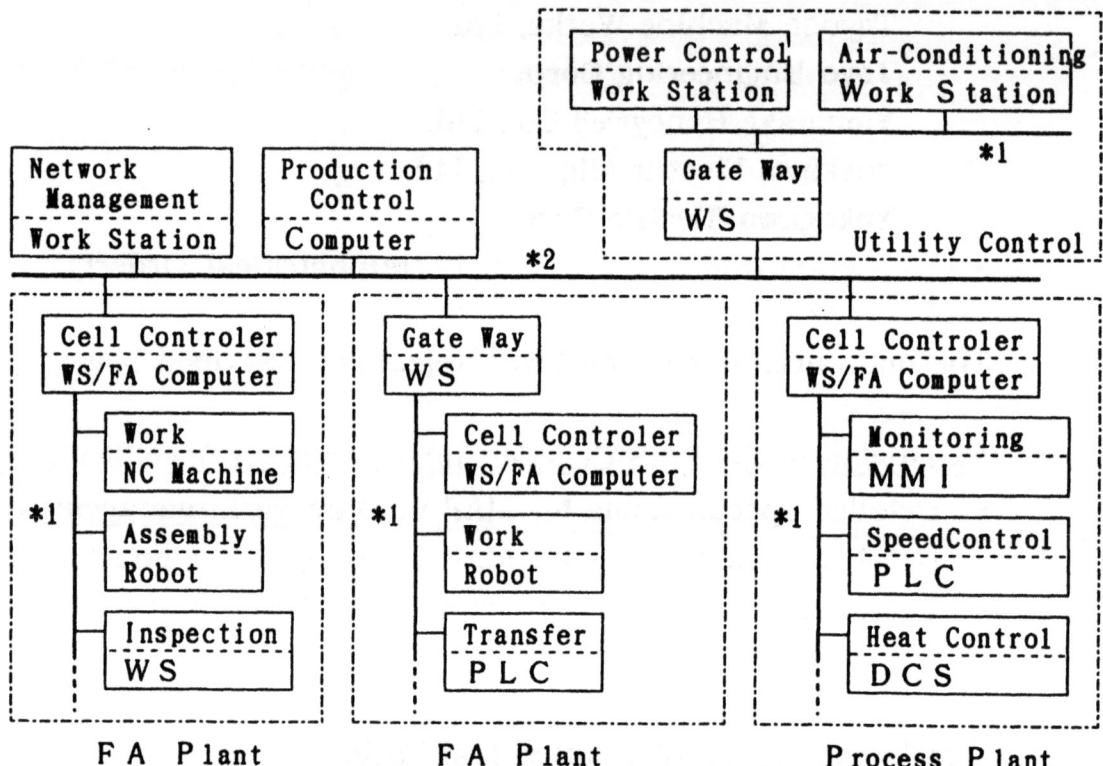

Note *1 F A I S C e l l - N e t (M i n i - M A P)
     *2 F u l l - M A P (T r u n k L i n e)

Fig. 4   F N E' 9 2   C o n c e p t i o n

EPRIT Project No 955/2617

C.N.M.A. - A EUROPEAN INITIATIVE
IN MANUFACTURING COMMUNICATIONS

S.C.H. Withnell

BAeCAM
Guild Centre Offices
Lords Walk
Preston PR1 1RE
Lancashire
United Kingdom

Abstract

The Communications Network for Manufacturing Applications project began work in January 1986. The CNMA mission is to to establish an open communications system to meet European industrial requirements, based upon emerging international standards. This paper reviews successes and achievements of the project team.

1.0 Introduction

Factory automation is characterised by the large variety of computers and controllers from many vendors in any real installation. Thus, installations often appear as autonomous "islands of automation".

The user is faced with two options for integrating his facilities. One option is to purchase computers and controllers from a single vendor and to use that vendor's proprietary protocols to achieve communication between the different devices. The user is then effectively tied to the products of that vendor and therefore may be restricted by a limited choice of devices. His alternative is to integrate computers and controllers from more than one vendor although he is then forced into considerable expenditure to achieve communication between numerous proprietary systems.

It is worth reviewing the benefits that can accrue from successful integration of manufacturing facilities.

  o Improved quality
  o Accurate and timely feedback of production data
  o Reduced response times to product changes

Of course, the bottom line is reduced costs. All these elements lead to a solid improvement in the competitive position of a manufacturer in the market place. An obstacle to the achievement of CIM has been the high cost of integration of heterogeneous CIM systems. In order to provide a cost effective solution for integration, a single communication system supported by many vendors is essential. The evolution towards Computer Integrated Manufacturing must therefore ensure that common standards are adopted by both vendors and users, especially in the key areas of Industrial Local Area Network communications and data management.

For some years, the Open Systems Interconnection or OSI has been proclaimed as the "enabling technology for CIM". Work in this area has been carried out under the auspices of the International Standards Organisation's seven-layer model for Open System Interconnection.

The most publicised examples of this work have been the MAP (Manufacturing Automation Protocols lead by General Motors) and the TOP (Technical and Office Protocols lead by Boeing) initiatives in the United States of America. This work has now been taken up by the standards bodies such as the National Institute for Science and Technology and the Corporation for Open Systems.

It was seen as vital for European industry that there should be a European programme to complement the other international initiatives in the field of manufacturing networks. This led to the formation in January 1986 of a project addressing a Communications Network for Manufacturing Applications, known as CNMA and supported by the Commission of the European Communities under the ESPRIT initiative.

British Aerospace is leading the powerful CNMA consortium to which many major European organizations have contributed. Since the inception of the project the following organizations have made significant contribution to the work:

USERS	VENDORS	RESEARCH INSTITUTES
Aeritalia	Bull	Fraunhofer - IITB
Aerospatiale	GEC	University of Porto
British Aerospace	ICL	University of Stuttgart - ISW
BMW	Nixdorf	
Magneti Marelli	Olivetti	
PSA Peugeot	Robotiker	
Renault	Siemens	

SYSTEMS HOUSES

Alcatel TITN
Comconsult

A specific aspect of the project has been to provide a conformance testing framework for validation of the vendors implementations. This has been achieved with the support of some additional organizations including SPAG, ACERLI and the Networking Centre.

CNMA's current requirements for conformance and interoperability test tools have been met by a complementary ESPRIT project known as TT-CNMA, lead by SPAG.

The user companies have great experience of Computer Integrated Manufacturing and brought to the CNMA project a clear understanding of the users' requirements and the limitations of current technology. The vendors recognise C.I.M. as a major market area for their products which include computers and the full range of controllers which modern automation demands.

It is important to recognise that this European wide combination of experience skills, and interests, all dedicated to a common goal, has been the major contributing factor towards the success of the CNMA programme.

Clearly, the work of the CNMA project has been a major undertaking. The commitment of the CNMA partners is based upon a firm belief in Open Systems communications as the enabling technology for C.I.M.

It must be stressed that CNMA is targeting the same final profile for manufacturing applications as the MAP programme. However, the MAP programme represents strongly North American needs. In order to provide balance, CNMA acts as a complementary programme representing the European users' requirement for multi-vendor interworking down to the level of device controllers. Within a manufacturing cell, large American industries use primarily a single vendor's controllers and thus proprietary communication systems. (Their main need is for a means of linking together a variety of different manufacturing cells.) However, many companies do not have the same level of investment, and demand multi-vendor working right down to the level of robots, machine tools, etc. Whereas file transfer can solve many problems at the factory backbone level, automation protocols are essential if open systems are to be achieved at this lower level.

At the outset of the project, the following project objectives were established:

o specification, implementation, validation, demonstration and promotion of standards and specifications for factory automation applications to ensure the development of standards suitable for European users ;

o compatibility with :
   MAP and TOP Specifications, and CEN/CENELEC supported standards, in Europe, to ensure that a single international profile is obtained.

o Promotion of European acceptance of standards, to encourage European vendors to adopt them.

o Encouraging the creation of validation centres, to facilitate the testing of vendors implementations.

In order to achieve these objectives the CNMA Consortium engaged in the following main activities :

o specification of CIM users' requirement ;

o selection of an unambiguous profile of communications standards to meet the users' requirements ;

o implementation of the profile on controllers and mini computers ;

o development of conformance test tools and conformance testing of the implementations ;

o validation and demonstration of the implementation in real production facilities ;   and

o promotion of the profile.

Although considerable effort has been expended to establish the standards for industrial LANs, there was insufficient information on the needs these standards are intended to satisfy. CNMA has enjoyed the advantage of including seven major industrial users in the project and has conducted detailed analyses of the subject. These cover a wide variety of requirements, in both batch manufacturing and the process control industry, including costs, time constraints, reconfiguration, redundancy, reliability, integrity and training.

The results of the studies are contained in reports which are available from the Commission of the European Communities.

The first activity in meeting the users requirement is the selection of a suitable, unambiguous profile of communications standards, incorporating both existing and emerging standards. This work complemented other activities leading to version 3.0 of the MAP and TOP specifications. The profile is documented in an Implementation Guide which becomes the specification for all CNMA implementations.

In order to prove that each implementation conforms to the standards specified, it is necessary to subject it to "conformance tests". The Fraunhofer Institute, the project's independent test organisation, obtained and developed the conformance testing tools and test procedures to verify the CNMA implementations in the first phase of the project, and worked with The Networking Centre, Spag Services and Acerli to develop testing tools and procedures for phases 2 and 3 of the project. In the current work, Fraunhofer -IITB participate in both CNMA and TT-CNMA providing a centre of excellence for conformance testing issues.

Perhaps the best validation of the CNMA implementations is by applying them to real production machinery provided by the users. These "pilot facilities" also ensure that the standards are appropriate to industrial use and provide the opportunity to demonstrate the project's achievements. The "pilot facilities" will be discussed later.

Finally, the CNMA consortium promotes the communications profiles to encourage their widespread acceptance. This activity is vital for achieving the ultimate goal of having a single unambiguous profile for manufacturing worldwide and is being performed by close liaison with standards bodies and by publicising the Implementation Guides. Releases of the Implementation Guide are available from the the European Commission.

2.0 Pilot demonstration facilities

The CNMA consortium has delivered a number of pilot demonstration facilities:-

   o a cell demonstrated at the Hannover Fair, West Germany,

   o three production facilities at:-

      a) BMW in Regensburg near Munich, West Germany
      b) British Aerospace near Preston, in England and
      c) Aeritalia in Turin, Italy.

   o two experimental facilities at:-

      a) the University of Stuttgart - ISW
      b) Renault, near Paris, France.

   o two production facilities at:-

      a) Aerospatiale, near Paris, France and
      b) Magneti Marelli, San Salvo, Italy

The Hanover demonstration provided the first public display of interworking of hardware and software in compliance with a CNMA Implementation Guide. It took place at the well known Hannover spring Fair in April 1987.

The main objective of this demonstration was to display true inter-working, using mini-computers and controllers, connected together through CNMA industrial local area networks. The cell was designed to be typical of manufacturing cells : it comprised a machining centre, a robot, a transporter and manual stations. With this system, machining of real parts was possible.

All this equipment was transported to Hanover for demonstration at the Intermatic exhibition as part of the world's largest Trade Fair. For eight days the cell was demonstrated and the aims and achievements of the CNMA project were explained to visitors from all over the world. This was the first ever demonstration of multi-vendor interworking using MMS, and the only such demonstration before

June 1988.

While exhibitions can be used to demonstrate new communications software, the most thorough validation is provided by using it in real production activities, in a competitive industry, where it must provide all the essential functions with very high reliability. The C.N.M.A. project has validated its communications software in three such environments... at BMW, British Aerospace and Aeritalia.

BMW's position in the highly competitive automotive industry, means that the company must stay at the leading edge of technological developments to produce ever greater efficiency. They were one of the first companies to recognise the benefits available from adopting standards for communications over these Networks and this led to their involvement in the CNMA project.

This CNMA production facility is at BMW's new factory in Regensburg, West Germany, where 3-series limousine and convertible cars are manufactured.

On the production line, car assembly tasks are controlled by well over one hundred PLC's which pass any disturbance or error reports to a Siemens Sicomp mini-computer. This information is then relayed to a Nixdorf Targon mini-computer over a Baseband LAN and a factory backbone Broadband LAN. Information regarding major disturbances is relayed over these networks immediately, using CNMA's MMS message transfer protocols, while less urgent information is transmitted at the end of the production shift, using MMS file transfer protocols.

This comprehensive maintenance scheme is the worlds first production application of M.M.S. It indicates a major industrial companies confidence in the software developed within the CNMA project, particularly since the facility involves just in time production. The result is, that since early 1988, this software has been used in the manufacture of cars which are a status symbol throughout the world.

CNMA's second production facility is at the British Aerospace factory in Samlesbury England, and machines 'D' shaped components for the leading edge of the A320 Airbus wing. This facility participated in the Enterprise Networking Event, and was the Events only true production facility. CNMA's participation in Enterprise has allowed the project's vast experience in communication software development to be applied to the evolution of the MAP specifications. The experience that CNMA partners gained through the Hanover Fair and the lessons learned, enabled them to achieve testing for Enterprise ahead of schedule.

The Precision Boring Machine, the centre piece of the facility, is served by an automated transporter and is supported by a number of manual process areas. The complexity of the control system allows the facility to be operated as a manned flexible manufacturing system for just-in time production.

As with the Hanover Fair demonstrator, CNMA has used more computers and controllers than necessary, simulating a large scale C.I.M. installation.

The vendors devices jointly control the Precision Boring Facility using CNMA software to communicate with each other.

An advanced feature of the CNMA communications software used in this facility is the Network Management Service. It allows the user to configure and monitor the performance of each layer of the communications software in each device. CNMA was further involved in Enterprise by providing a major part of the required MAP test system, through the third phase of the project, named CNMA Conformance Testing or C.C.T. These test tools, which are necessary to ensure inter-operability of devices, enabled the Enterprise Networking Event to proceed. This conformance tool set is now marketed world wide, by SPAG-CCT, a Belgian company established specifically for this purpose.

CNMA's involvement in this Event, demonstrates that the same software standards are being adopted on both sides of the Atlantic. This is a major step towards enabling manufacturers to integrate devices from different suppliers, European or American to create a Computer Integrated Manufacturing environment.

The ability to transfer manufacturing data between companies over a computer network will become an increasingly important feature in advanced industries. For example, companies frequently wish to send updated manufacturing data to a sub-contractor. In November 1988 at the Aeritalia factory in Turin, CNMA commissioned its third production facility, integrating mini-computers from Olivetti and Bull. This facility is used in the manufacture of aircraft wire harnesses. As an enhancement to this pilot facility, an X.25 network connection was established between the British Aerospace pilot and Aeritalia. This Wide Area Network link allowed the transmission of wire harness manufacturing data from British Aerospace to Aeritalia, demonstrating communication with a sub-contractor.

More recently, a highly sophisticated experimental facility has been commissioned at the University of Stuttgart. Within the facility, ISW has used equipment from Bull, GEC, Nixdorf, Olivetti, Robotiker and Siemens along with applications software from Alcatel-TITN to build an impressive demonstration. The ISW facility consists of two independent cells, one is designed to manufacture bodies for hydraulic valves. This cell features a turning centre, a boring and milling centre, a linear portal robot and a pallet store. This cell is fully automated. The second cell consists of a manually loaded 5-axis milling machine and is connected by a local area network to a component design facility consisting of CAD and NC programming systems. Once part programs have been created, they can be downloaded to the NC controller and started by the cell controller.

The next facility to be commissioned will be at a Renault factory in Boulogne Billancourt, France. This pilot provides a test and demonstration facility for Network Administration systems. In large networks, automated facilities are required to handle faults, tune the network for optimum performance and manage re-configuration of the network. Suitable software for this is provided within the network management function. The facility consists of a conference room connected to a test laboratory. Three demonstrations are planned:-

   o A simulation of the Aerospatiale pilot.
   o Control of video equipment and the communication of video images.
   o Management of faults achieved by use of a fault injector for simulation of typical "real world" faults.

In combination, the facility will demonstrate the ability of the network administration system to tune and reconfigure the network and to diagnose and manage fault conditions.

At this stage in the project a turning point exists, namely, a move from looking for technical confidence in open systems solutions to assessing the business benefits that can be achieved using stable OSI implementations.

The first facilities which can be assessed in this way are:

   o Aerospatiale, near Paris, France.
   o Magneti Marelli, San Salvo, Italy.

The Aerospatiale facility uses CNMA implementations for the integration of a machine shop. This machine shop is used for the production of prototype missile components. There are nine machine tools in the facility and the objectives of the facility are improvements in quality, flexibility and delivery times. The CNMA communications software is used to integrate four applications. These applications control all aspects of cell control such as maintenance management, shop management, scheduling and transport control. The aerospatiale pilot will become operational during December 1990.

At Magneti Marelli, the CNMA software is installed in the final section of an alternator production line. Three primary functions are performed within the facility. These include shop floor monitoring of machine productivity and performance, a tracking system traces all work in progress through the shop and a diagnostic system collects detailed information on machine status. This manufacturing facility will become operational during December 1990.
These facilities use the very latest implementations of International Standards, the core services of which are now stable.

3.0 CNMA communications profile

Communication in CNMA is based on the Open Systems Interconnection basic reference model (OSI/RM), which is defined by ISO in IS7498. It defines a framework for communications, that is the services to be provided to application processes, the breakdown of the communications software into 7 layers and the split of services between the layers. The model is known as the "OSI seven layer model".

Each service requires protocols - specified interactions - between the two systems. Definitions of the services and protocols in each layer are provided in individual standards documents. However, a given service can be provided by a number of different protocol combinations in the lower layers. Hence, an additional document is required to identify the exact protocols used at each layer. Such a selection of protocols is known as as a "profile".

The CNMA Implementation Guides defines the communications profile chosen for use in the project. They represent a considerable amount of work by the participating companies and their publication remains a primary function of the consortium.

The main purpose of CNMA is to focus its research upon application layer (layers 6 and 7) issues. This was also a major topic in MAP developments and has been addressed in MAP V3.0. CNMA has contributed to the MAP evolution through its research and implementation work in this area, so aiding the definition of a single international profile of communications for use in the manufacturing environment. CNMA is making strenuous efforts to maintain compatibility with MAP, and also with the standards supported in Europe via such groups as SPAG, the vendor Standards Promotion and Application Group, and through the standards bodies such as CEN/CENELEC.

The structure of the Implementation Guide is as follows:-

Volume 1: Transport Profile The purpose of the transport profile is to address the issues of network connection, network addressing and reliable data transfer. This volume encompasses the lower four layers of the ISO model.

Volume 2: Application Profile The application profile contains mechanisms for the control and passing of messages between applications. Specific protocols and services such as MMS, FTAM and Directory Services are addressed.

Volume 3: Application Interfaces By standardizing the interface to the communications software, the portability of user applications is greatly improved. The FTAM and MMS interfaces are described here.
Addendum 1: Network Management Application During the current phase of work, CNMA has expanded much effort in the area of network administration applications.

Addendum 2: Technical Reports This addendum is a analysis of certain CNMA related topics such as Fieldbus, Remote Database Access and Higher Performance Architecture.

CNMA Implementation Guides devote a chapter to each layer or service.

For layers 1 and 2, CNMA initially uses Local Area Networks (LAN's) and acknowledges the benefit to users of providing a choice of LAN type. This allows a user to choose a type based on : cost; performance; installed base; maintainability; etc. Three options are specified, and all have been successfully utilized within the project.

MAP has initially opted only for the broadband technology and extended this to cover carrierband, with token bus access, but studies in Europe showed that 98% of LAN installations have opted for baseband technology. CNMA's choice of baseband LAN is the same as that supported by TOP.

In CNMA, layers 3 to 5 are designed to conform with MAP. These layers are more stable than layers 6 and 7, and are therefore defined as 'background' for the project. This enables work to be concentrated on the other upper layers.

For layer 6 (the presentation layer), which was absent in MAP V2.1, CNMA specified the use of a kernal subset. Layer 7 protocols were covered in a number of chapters. The first defined the Association Control Service Elements known as ACSE, which used the latest ISO draft international standard protocol to provide association control.

A chapter for layer 7 protocols defined the Manufacturing Message Specification - MMS, which is a protocol for passing messages between computers, numerical controllers and programmable logic controllers. This was the greatest area of activity in the project.

It is not necessary or desirable for vendors to implement the complete MMS service, so subsets were selected to provide the functionality required for the demonstrators. Typical services were selected as follows:-

   o Up and down-load programs to PLC's and NC's
   o Report change of status from NC's and PLC's
   o Start and stop program execution in PLC's and NC's
   o Request status information from NC's and PLC's

In all of these services, the NC or PLC interacts with a computer. For the definitive list of supported functions, the Implementation Guide should be consulted directly.

Another chapter in the layer 7 group presents FTAM, currently file transfer and management of file structures, is supported.

A further chapter for layer 7 covers network management. This is a protocol which permits a remote device (manager) to access communications related attributes (counters/timers) in other devices (agents). Using this protocol a manager can monitor and influence network performance, configuration etc.

Finally, a chapter in this series covers directory services. The protocol and resultant services, allows devices to interrogate a remote directory database. This can be used for example, to establish the address of an application it may wish to transfer a file to.

The CNMA profile is not yet firm, but evolves as the upper layer standards mature. The latest issue of the Implementation Guide produced by ESPRIT Project 2617 is version 4.1 and incorporates the latest ISO specifications including :

  o MMS and MMSI
  o FTAM and FTAMI (File Transfer, Access and Management),
  o Directory Services, and
  o Network Management

In the current work, network administration is seen as a vitally important service for large computer networks. Consequently, network administration is an interesting market area for the vendors to exploit. The tasks that the CNMA network administration system is designed to perform include:-

  o Configuration management
  o Performance management
  o Fault management

Further, by use of knowledge based techniques, the network administration system can carry out these task in an autonomous manner.

The business objectives are to reduce down time, optimise network performance and simplify the maintenance and support function. These functions are described in Addendum 1 and are Network Management applications rather than protocols.

4.0 Conformance testing

If industry is to reap the benefits of open systems it is essential that conformance tests are established worldwide which provide consistent and uniform results. This is a necessary step towards ensuring interworking of products from different suppliers.

In the first phase of the project the Fraunhofer Institute, an independent testing organisation, obtained and developed the conformance test tools and procedures to verify the CNMA implementations. For testing MMS and ACSE, they developed a test system on a mini-computer system programmed in 'C' and running under UNIX System V.2. The tools developed in this environment are highly portable, permitting future delivery to vendor or user sites, and to other test institutions and developers. This was the first such tool developed for the MMS application protocol which is the main feature of MAP 3.0.

For testing the lower layers of CNMA implementations for the Hannover Fair demonstration in April 1987, Fraunhofer obtained the MAP 2.1 test-bed from the ITI in Michigan. These tools were the only ones recognised by the US MAP User Group for MAP 2.1.

During the pre-staging for the Hannover Fair, the test tools were used to great effect, testing the vendors' new implementations. The outstanding success of the demonstration is evidence of the quality of the test tools developed so far.

The urgent requirement for test tools for MAP and TOP at ENE '88i provided a rare opportunity for Europe to establish and own a significant test suite by bringing together the necessary resources under the ESPRIT programme. Besides the existing CNMA partners, SPAG Services, the Fraunhofer Institute, The Networking Centre and ACERLI became involved. Agreements with the Corporation for Open Systems (COS) in the USA have been made to ensure the widest possible acceptance of these conformance tests.

The third phase of the project, CNMA Conformance Testing developed conformance tests for MMS, Network Management, Directory Services, Bridges, Routers, End System/Intermediate System Protocol and Logical Link Control Class 3. These tests were supplied to the Enterprise staging areas. SPAG established a belgian company, SPAG-CCT SA, who have successfully completed product development and the tools are now marketed world wide. Further work has been sponsored under ESPRIT II in the form of EP 2292, Testing Technology for CNMA and is known as TT-CNMA. TT-CNMA is co-ordinated by SPAG, and is furthering the development of conformance test tools. TT-CNMA is also leading the development of Interoperability test tools, which are used to analyse the interoperability characteristics of conformance tested devices. These "IOP" tools can also perform load testing and log protocol interactions. All these aspects enable the integration team to establish confidence in the communications sub-system before installing applications software.

5.0 Conclusion

In conclusion, the major achievements of the project can be reviewed as follows:-

   o Four Implementation Guides have been produced defining a
   communications profile to allow MMS automation protocols, FTAM,
   Network Management and Directory Services protocols to be exchanged
   between multi-vendor systems of mini-computers and programmable
   devices.

   o Standardized user interfaces have been developed for MMS and FTAM to
   increase portability of application software.

   oNetwork management applications software has been developed to allow
   superlative control of large networks to be achieved.

   o Multi-vendor control using CNMA communications software has been
   successfully demonstrated at the 1987 Hannover Fair, the Enterprise
   Networking Event, and in real production environments throughout
   Europe. A number of world "firsts" have also been achieved during these

activities.

o CNMA is liaising with standards bodies and is having an impact on the final, single communications profile, thanks to its experience in implementation and validation of the CNMA profile.

o The project has brought many European vendors closer to the final standard.

o A comprehensive set of MAP 3.0 test tools are now marketed world wide by SPAG-CCT, a commercial venture launched specifically for the marketing of the conformance test tools.

o CNMA vendors such as BULL, GEC and Siemens now market OSI based products. These MAP compatible products include mini-computer interfaces, programmable logic controllers, gateways and bridges.

The major impact the project has had on emerging standards is already being built upon with major new work extending into the 1990's. This new work will be funded under an ESPRIT II contract EP 5104. This new contract will ensure that ESPRIT CNMA continues to make a major contribution to the lower cost CIM facilities that manufacturing industry so urgently needs.

6.0 References

(1) CNMA Implementation Guide 4.0

(2) Industrial Local Area Networks: User's Needs

(3) CNMA Strategy Document

(These documents are available from the ESPRIT-CIM project office)

IMPLEMENTATION STRATEGIES, VENDOR PERSPECTIVES

CIM APPLICATIONS: A SYSTEM INTEGRATOR'S VIEW

Adriano De Luca
Lelia Giovani
Franco Marra

Syntax Factory Automation
Olivetti Information Services
Via Vela 27 - 10128 Torino
Italy

Summary

By way of introduction, CIM market requirements are described, as perceived by a System Integrator (Syntax Factory Automation, a company of the Olivetti Information Services Group).
Solutions and approaches are then described, by way of a case study, in terms of their ability to properly satisfy market requirements, and major pitfalls that cannot be avoided with current technologies are taken in account.
The CNMA technology is then analysed as a possible solution, pointing out its benefits and areas of possible improvements.
In conclusion, comments will be made about the overall potential of CNMA to become a standard environment for System Integrators.

1.Introduction

Syntax Factory Automation (a company of Olivetti Information Services - in the following referred as SyFA) is targeted to operate in the system integration Italian and European market.

The value of the Italian system integration market is about 300 M$ (1990) with an annual growth rate of 15%, 80% of the market being represented by companies with a turnover greater than 80 M$. 50 % of big companies (with a turnover greater than 400 M$) represent a market close to saturation (having already installed solutions), while medium-high companies (with a turnover ranging from 80 up to 400 M$) offer good (even if difficult) market opportunities. Also mid-sized companies (with a turnover ranging from 40 up to 80 M$) offer good market opportunities, but they often do not have expertise in handling system integration solutions. Finally, small enterprises (with a turnover less than 40 M$) are a virgin territory (source: IDC Italia 1989; exchange rate: 1250 liras per dollar).

This specific market is served by SyFA by means of a quite complete offering:
- organisation consulting
- system design (algorithms, software architecture)
- system implementation (software development, integration, start up, maintenance and support)
- proprietary software and solutions (plant simulators, resource and materials schedulers)

As an answer to the increasing competition, the SyFA offer is evolving from a "work order" approach to a "near product" style in order to improve quality and to cut development risks.

Figure 1 gives an idea of current SyFA performances.

1989 turnover	**12.5 M$**
1989 profits (before taxes)	**1.8 M$**
1990 turnover (budget)	**14.0 M$**
people	**55 (48 engineers)**

fig.1: SyFA main figures

2. Major market requirements

CIM market is expressing a certain number of requirements; of those, in SyFA opinion, the "stronger" ones are:

1) investments should result effective in their return. Entrepreneurs, or managers, who usually have to deal with strong competition in their specific market, are very sensitive to effectiveness of their company investments. On the other hand, CIM solutions often are very expensive.

2) Partners rather than suppliers are searched. CIM solutions could be very complex in their design, implementation and usage. So Customers often need expertise they have not, support and training. Design phases involving organisation problems should be carried out together, along with installation and start up, while implementation should often be verified in partnership.

3) Systems should be available. When systems operate, minimum time for maintenance and, if it is possible, no down time is allowed

4) Solutions should be designed to be very flexible. This need comes up from the fact that from initial specification to final installation and startup sometime a considerable amount of time could be spent: in complex organisations things may change during that period. Moreover a good solution should be able to adapt itself to new arising Customer's needs and modifications during its operating life

5) Different systems should be integrated. A complex organisation requires solutions based on hierarchical models, each level of the hierarchy requiring the proper computing platform, ranging from host computers down to PC's. Moreover, Customer's policy could state to maintain independence from a specific IT supplier. In any case, different computers and different software could be required to cooperate in final solutions. Especially in CIM applications, as better said in the following, existing solutions, representing "integration islands", are often requested to be gracefully integrated in the overall design: this point more and more stresses the importance of the ability to connect, use and

integrate etherogeneous computing platforms.
6) Systems should be friendly. Easiness in using, in maintaining and in evolving them is mandatory.
7) Respect of existing Customer internal standards. These standards are very often defined for each of the levels of the solution hierarchy - mainframes, minicomputers or PCs, PLCs, networking, LANs, CAD and CAM machinery, NCs and so on - see point 5) above.

Up to now, only "generic" CIM requirements have been described. But, to be more specific in our speech, Customers can be thought as belonging to two main classes, each of those having its peculiar needs: "from the grass" Customers and "existing plants" ones.
Strangely enough, "existing plants" Customer's problems usually are a bit more difficult to be solved. In fact those Customers are characterised by additional requirements:

8) existing plants and machinery should be gracefully integrated with the new ones. Note that, especially in the case of small and mid-sized companies, existing solutions are very often not automated; nevertheless, those working installations should be preserved in order to match automation costs with the investment capacity of the Customer. Example of those are existing CAD and CAM machines, or trucks or manually served racks and so on.
9) computers have to exchange information with other computers already installed and working well. MRP stock control and administrative applications running on mainframes, PC's in control of existing parts of the plant, and so on, often are existing systems to work with. Note that those automation islands often are based on solutions dictated by major market players, establishing in such a way a number of de facto standards widely disseminated within the CIM scenario.

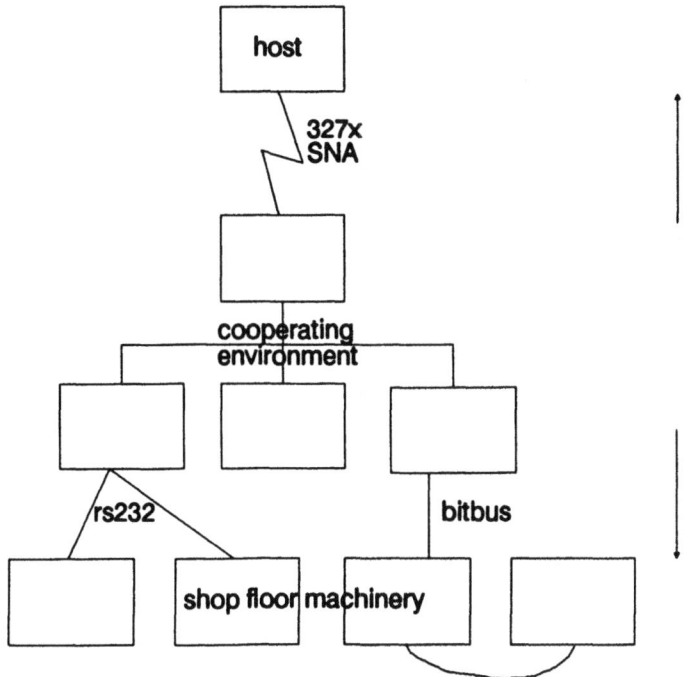

fig. 2 : connectivity in different computing levels
and within the same level

Last but not least, requirements themselves are evolving in time, requiring more and more integration through better connection abilities. For instance, to put together engineering and production support in a common product design philosophy, requires the active cooperation of CAD and CAM machinery, administrative packages, quality control, documentation and information retrieval systems, along with planning control software and the ability to download partprograms and to control shop floor machines.

In few words, the market seems to require effectiveness in investments, partnership in problem solving, high availability and maintainability of solutions along with flexibility and robustness, and the ability to have platform independent solutions, able to connect and integrated etherogeneous software, packages, computers and machinery, some of those already existing and running. "Connectivity" really is a keyword in CIM world!

3.The Syntax Factory Automation Response

The above points are very strictly related to the subject of this paper. All those requirements have some relationships with concepts like "etherogeneous computing platforms", or "portability", or "flexibility", or "robustness", or "fault tolerance", and, above all, like "connectivity". These terms, in turn, can be regarded as keywords in defining requirements for software design and development.

Let's describe how SyFA is currently approaching the overall problem. Software is designed to be portable adopting standard environments where available, or designing proper interfaces implemented upon the most common operating systems and communication environments.

Flexibility and robustness is gained through a design approach which adopts advanced tools, a strong modularisation and high level interfaces. In other words, object oriented and 4th generation languages, SQL interfaces and a configurable library of software modules are in use.

Fault tolerance is reached adopting redundant solutions: they can be implemented by using LAN technology or fully fault tolerant high end computers.

Connectivity with hosts, is usually achieved by means of 3270 emulator, while connectivity with shop floor is based on a number of specific protocols. On the other hand, computers belonging to the same "level" speak to each other via native IT suppliers solutions, usually LANs based on the most common wiring techniques. In any case, a lack of a "clean" architecture and approach is felt. Moreover, both in case of local and remote connections, network management and control is very often still an open problem.

In the next chapter, a real example, describing, from a design point of view, the SyFA Material Handling offer, will better support our thesis. Starting once again from the market, software architecture requirements are derived, leading to the definition of the SyFA native software environment. That platform, will be used, in the rest of the paper, to analyse the CNMA approach advantages and disadvantages.

4.The SYFA Material Handling Offer: a real example

Italian market requires a material handling offer organised in three levels. The low level is represented by a little warehouse control ranging in price up to 25 k$, mid-sized one is made by a system able to have a certain level of redundancy, with a end user price up to 70 k$, while high end products must have absolute fault tolerant characteristics; in this case price is up to 300 k$ or more. Mid-sized and low end systems, moreover are required to be integrated with already existing low technology plants and transportation system (i.e. man operated trucks and racks), while high end will usually be in control of fully automated transport systems.

From the IT suppliers offer, PCs for the low end, PCs or minicomputers along with LANs or proprietary networks for the mid-sized solution, and fully fault tolerant systems for the high end segment can be adopted as computing platforms. Well known trade marks are: MS-DOS, Xenix, Unix, VMS, OS/2, LAN MANAGER, DECNET, mVAX, CPS and Stratanet. As a rule, Customer's internal standards determine the computing platform choice.
The system Integrator's problem is therefore to give a common solution for all three offering levels, in order to build up a "product line", saving development and design costs, running on top of the "Customer preferred" computing platforms.

Design has been carried out by first clearly defining two main modules: application and environment. Application will be customer sensitive but independent from computing platforms; on the other hand, the implementation of the environment will be platform dependent. Interfaces clearly separate the application from the environment.

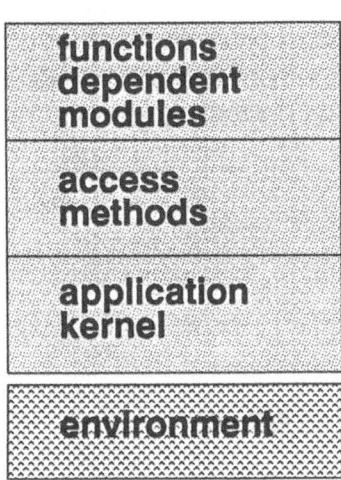

fig. 3: SyFA platform general architecture

The application module has been organised in terms of an "application kernel", that can be reused in any installation, and a number of libraries to match specific Customer's needs. An object oriented language helps in the design and in the proper data representation, making the code easy to be changed and prone to evolution and maintenance.

While the application architecture solves the specific class of material handling problems, the environment architecture has been designed to solve the general problem of the portability and connectivity of any application (Material Handling, Production Control, Engineering support and so on) on top of the most common computing platforms and networks. Portability and connectivity are matter of interfaces. So in the following, the basic SyFA environment, will be described in terms of its in

Interfaces have been chosen, where available, from the market. Access to data is guaranteed via SQL. The fact that SQL is supported by major DBMS system suppliers allows portability, and makes it pretty easy to choose a DBMS by taking in account actual system performances.
Process to process communication is supported, using a home defined interface (IPC). This interface is designed to support inter process communication, processes being subdivided in two classes, requesters and servers. Current implementations of this interface run on DECNET, LANMANAGER and TCP/IP, "raw" RS232, and locally on the same machine, allowing distribution of processing (that is the normal case, in mid-sized offering). Moreover the same interface is used to communicate with host and with plant. Special implementations of this interface use 3270 emulation, in SNA environment, and support a number of PLC and NC protocols using RS232 wiring. Properly configuring the system, it is possible to have a "secure" communication protocol. Messages can be logged, and cancelled only when sender is sure they have been properly received by the "receiver". A process-to-process commit tool is supported by the IPC implementation, this ability resulting, in case of bad functions, in a useful tool to guarantee system transactionality. No other tool (network control etc.), a part those which come with native networks, are used, nor special international standardisation of messages and protocols have been adopted.
Other interfaces, less important from the point of view of this paper, are those regarding Error, Messages and Log management, process management and user interface handling.

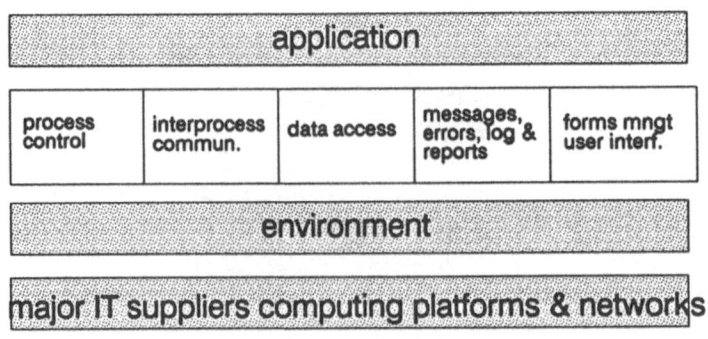

fig. 4: SyFA native environment: major interfaces

5. IPC: A solution for networking and interprocess communication

In the previous chapter, the general architecture of SyFA native environment has been described. Let us now better analyse the specific networking and interprocess communication interface (IPC), along with its major advantages and pitfalls.

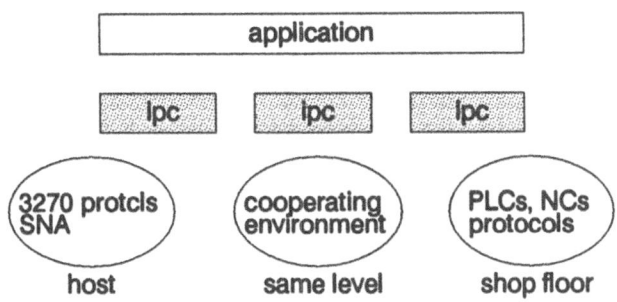

fig. 5: IPC and its implementations

By using IPC, it is possible to write applications independently from wiring, locality, operating systems and network environments and with logical level naming conventions. Moreover, a "secure" protocol, and a commit feature, guarantee, where necessary, that messages are properly received, and that the system transactionality is supported. IPC has been implemented locally on the same machine on a number of operating system, and on top of the most common local protocol and LANs, and special implementations of it regards communication with host via 3270 protocols and with shop floor machinery via RS232 cabling: both administrative environments and shop floor installations are seen, by applications, like servers, providing to the system a very nice uniformity of approach. In particular, those implementations are respective of the market realities, and match very well with existing solutions and automation islands.

On the other hand, messages are still application dependent, and no general tool is supplied to build them. Messages to host, are 3270 transaction masks, and messages to shop floor follow machinery conventions. Application programmes should be aware of this aspect, reducing so the overall portability in a significant measure. The absence of a way of classifying classes of messages, and of a tool to build them, is of a certain importance in new applications, where many etherogeneous machines (CAD, CAM, Quality control and planning packages etc.) are requested to be integrated within the overall CIM solution. No general service in supporting application independence from the transferring syntax point of view is provided. As in the case of messages, applications are requested to be aware of buffer dimensions and other details, limiting so their generality. Moreover, actions and logic by which communication takes place are explicitly implemented in the applications. Network control, rerouting abilities, tools to understand "what's cooking on the net" are those which come with the IT suppliers native network environments.

In other words, SyFA current solution works very well for current implementations, but the lack of generality and control could result in special developments for future projects, and in a possible poorness in redundancy for what the network is involved. No "added value" is built to what is supplied by standard network packages (LAN MANAGER, DECNET, TCP-IP, SNA etc.) in terms of network control. On the other hand, IPC is very well designed to match with existing standard, and nicely supports transactionality.

6. CNMA: An attractive solution

For sure, the CNMA approach is designed to inherit from ISO/OSI all the advantages due to the clean architectural approach and to a clear definition of standard services and options (profile). Major problems of SyFA IPC, as described in the previous chapter, are due to the lack of such a clean architecture. But, as far as this paper is concerned, it is useful to stress more some peculiar CNMA advantages, at least as they are perceived by SyFA. If we analyse the CNMA pile starting from the bottom, we can find the following major plusses:
- the ability to have installations based on a range of different wiring techniques: CSMA/CD and TOKEN BUS in broadband or in carrier band are available, allowing the most suitable net topology to be chosen (branching tree);
- the fact that CNMA is an ISO standard for transport and session: now a day all major IT suppliers adhere to those standards, allowing so multivendor solutions;
- in particular, the transport class and network layer within CNMA profile, supports a large spectrum of network topologies, allowing such features like rerouting, and traffic optimisation and flow control (that are typical of MAP);
- a large number of general purpose "application services". Among these, a special interest is due to NMT (network management), a service in control of performances, faults and configuration. This point seems to be important to let solutions face up with requirements regarding system availability and effectiveness (tuning, performance etc).
- specific CIM oriented services, like MMS (Manufacturing Message Specification) and DS (Directory service) allowing general classes of messages to be defined, and naming definition.
- built-in message services, like PLC1, PLC2, NC1, NC2, CC supporting major PLC and NC suppliers' protocols, along with the ability to effectively build up other classes of messages in control of other shop floor devices.
- easy integrability with other platforms or services designed to match with other specific requirements. For example, the ability to run on the same network architecture applications based on X400, provides a straightforward way to put together CIM solutions with office automation environments.

In summary, CNMA architecture seems to be very appropriate to guarantee a complete application independence, from protocols, transferring syntax, communication logic, computing platforms and networks point view. Moreover network control is fully provided, in rerouting, in traffic optimisation and in network administration. Last but not least CNMA is an "open system", able to integrate with other ISO/OSI environments. An application program written using

CNMA can be transferred onto other systems adopting the same technology and CNMA application programs written by different suppliers can be easily integrated.

7.CNMA possible areas of improvement

As told before, the general design and the resulting architecture of CNMA seem to be very attractive for companies, like SyFA, that every day have to face up with the complex reality of the CIM market.

On the other hand, CNMA approach seems to collect very well new trends in IT technology: very often, modern computing platforms are based on Unix, or are Unix derived, and many and many applications (see for example CAD packages) run or are predicted to run on top of Unix environments. Moreover, not hierarchic, X25 based network architectures allow a more flexible services distribution, with respect to classic architectures. In few words, CNMA seems to be very nicely conceived to match near future requirements.

Unfortunately, the Italian market, as perceived by SyFA, especially in the case of the "existing plant" Customer segment, is characterised by the presence of hierarchic network architectures, and by computing platforms other that those belonging to the Unix family. Well known examples are IBM and DEC computing platforms and network architectures. In fact, at least in Italy, many of the administrative applications run on typical IBM environments: CICS, TSO, COPICS, 327x connections, are well known examples of what market is using in their current solutions. Moreover, for a number of reasons, DEC is in Italy a well appreciated IT supplier for Factory Automation applications, thanks to the completeness of its base environment and the real time capability of its computing platform.

Another facet of the problem is represented by the way shop floor machinery very often is physically connected, in real cases, to control computers. "Raw" RS232 is used, due to the simplicity and the cost effectiveness of the connection.

If we recall what has been said in describing the SyFA platform, we can see that the SyFA and CNMA approaches are pretty complementary. Also if a certain level of portability of applications and of fault tolerance is provided by SyFA platform, the overall functionality is not so complete like those furnished by CNMA; moreover the approach is not so well structured as CNMA one. SyFA approach was born on the basis of a day by day experience, but it results to be very effective in integration with existing realities and in support to transactionality. By a practical point of view, it is possible to connect solution with the most widely diffused network architectures, both at host and at shop floor level, in a secure way.

So we think that CNMA architecture could be improved, by adopting some functions found very useful in SyFA experience. The ability to integrate with hierarchic architectures, the usage of platforms other than those Unix derived, the adoption of "poor cooking" wiring techniques, the support to transactionality, could make CNMA a very attractive and complete network platform for system integrators. Note that the inability to cope with existing installations and architectures or to be portable to different computing platforms could be regarded as the most serious business barrier inhibiting the take up of Open Systems Technology.

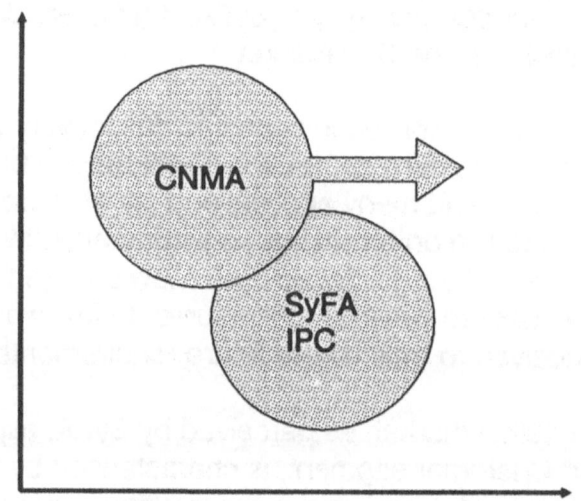

- adherence to existing standards
- effectiveness in current installations

fig. 6: CNMA and SyFA IPC functional areas
a possible CNMA evolution

8.Conclusions

Market requirements have been analysed, as perceived by a System Integrator, active in the Italian and European scenario. From these, major design requirements have been derived, and how they are used in actual implementation by SyFA has been shown, using a real example. Current implementation clearly shows a number of problems or limitations in satisfying requirements, especially in terms of portability, redundancy and network management and control. On the other hand CNMA seems to be very well designed and prone to fully satisfy market needs, especially in terms of evolution, portability, redundancy and standardisation. No special technical suggestions have been given to improve CNMA functions, but, rather, comparing it with current SyFA network technology, CNMA has been found quite satisfactory. On the other hand, CNMA could be improved to allow use of market significant computing platforms (others than Unix), of integration with other network architecture and of "low cooking", but already stated, connection technologies.

As a company, we plan to work closely to the CNMA project, to take advantage of CNMA progresses and of the competitive advantages that will result in bringing CNMA technology to the market.

9.Acknowledgments

Special thanks to Luigi Rispoli and Tommaso Ricci, of Olivetti Information Services Ricerca, for the opportunity they offered to contribute this paper, and for the support in writing it, and to
Fernarda Ferrari and Arvind Narang, of Syntax Factory Automation, who conceived and designed most of the software architecture described in this paper and the SyFA IPC interface.

CPS is a trademark of Stratus Corporation
DECNET is a trademark of Digital Equipment Corporation
LANMANAGER is a trademark of Microsoft Corporation
mVAX is a trademark of Digital Equipment Corporation
MS-DOS is a trademark of Microsoft Corporation
OS/2 is a trademark of Microsoft Corporation
SNA is a trademark of IBM Corporation
Stratanet is a trademark of Stratus Corporation
Xenix is a trademark of Microsoft Corporation
Unix is a trademark of AT&T Corporation

OPEN SYSTEMS AND SMEs

ARE THEY COMPATIBLE?

Sharron Burgmeier

Kewill Systems Plc,
Ashley House,
20-32 Church Street,
Walton-on-Thames,
KT12 2QS
U.K.

1. Summary

Mention Open Systems Interconnection (OSI) to most Small Manufacturing Enterprises (SMEs) and you will either draw a blank stare or an informed shrug accompanied by the comment "Interesting, I suppose, if you're GM, but not very relevant to me". Further probing will bring comments about OSI being too expensive and too complex a solution to the perceived simpler communication needs of the SME.

In this paper, we consider the truth of those statements. We examine the financial implications of OSI in the SME environment and draw some conclusions about the broader issue of standards in operating systems, relational data bases, user interfaces, and languages as well as in communications. This suite of "standards" will be referred to as Open Systems within which OSI is a component.

It is clear that end users remain very sceptical about the true commitment of major vendors to Open Systems. Many users believe that vendors will continue to exploit their differences as competitive advantages and that they will continue to develop their own versions of "standards". This wide-spread scepticism is retarding user take-up of Open Systems. After all, how can they commit to such systems if the vendors cannot?

Our conclusion is that Open Systems today is a win for applications vendors who can facilitate the porting and subsequent product offerings of software across multi vendor platforms. OSI specifically has the potential to be a win in the communications areas as users want better communication across multi-platforms.

However, as long as hardware vendors remain ambivalent about Open Systems, as long as they continue to give the appearance of paying only lip service to developing products which are conformant to major standards, so long as the controversies persist which prevent agreement of standards in operating systems, languages and data bases, SMEs will not be able to realize the full potential of Open Systems.

2. Introduction

For two years, Kewill has been preaching the gospel of Open Systems to our user community, made up of 6000 Small Manufacturing Enterprises (SMEs) and close to 300 larger manufacturing concerns. During that time, we have been developing CIM products which are designed to conform to the major national and international standards, this in an attempt to bring the benefits of those standards to our end users. Our difficulty has been in determining what those benefits are, which standards were important and how best to design to those which are important, and most difficult of all, determining what the hardware vendors really intended doing about standards.

The hardware and software vendors speak with great intensity about their commitment to Open Systems; they fund work on standards and join various and sometimes competing standards "groups". However, they continue to bring out proprietary products and indeed give the appearance of focusing their attention on such products. Perhaps it is asking too much of them to expect anything else. They are in competition and their differences are their marketing edge.
Given this, how does an application vendor build products with Open Systems as a design goal? Kewill began by reviewing the state of standards.

Standards Overview

As application vendors currently developing next generation product offerings, Kewill has attempted to cut through the jargon and hype of OPEN SYSTEMS to see what is there of value for SMEs. We looked at POSIX and first examined the IEEE standard for Portable Operating System Interface for Computer Environments. This standard is based on the UNIX* Operating System document to support application portability at the source level.

The 1003.1 standard is the first of the group of proposed standards known collectively as POSIX. It is intended to compliment other standards that in total would provide a comprehensive Open Systems Environment. Included in this group are C language standards, Networking Standards, and in particular, the standards for the OSI Model and data base standards, specifically the SQL and NDL standards.

We found that while the major hardware vendors stated their commitment to POSIX and funded the major standards groups working on "finalization of standards", they continue to put the majority of their focus and development attention on furthering their proprietary products. We know of a vendor who participates in every major standards event while privately telling Kewill with laughter that POSIX is not important to them and should not be to us. Is this just one not very astute marketing man's way of getting Kewill to focus on product offerings available now or does it reflect the "real" objective of that vendor?

Incidentally, singling out one vendor would not be at all fair. Kewill has been visited by representatives of almost every major hardware vendor one can name. The vendors are united in seeing that they must be solutions providers selling applications in order to sell more hardware. In support of this goal, they are all looking for next generation CIM product offerings to allow them to compete in the manufacturing market sector, a sector they see as showing high growth for the next decade. Almost to a man, they state their commitment to Open Systems citing their financial support for this and that standards group. However, in the next breath, they talk about "reality", the need to offer something different, to differentiate themselves in the marketplace; to have something better to offer their customers. It is hard to disagree with them on a commercial basis.

Having found a mixed state of affairs when talking about the broad "standard" represented by POSIX we decided to look at something much more focused, the C language standard. Surely this standard which had been around for some time and which represented a mature language would prove to be fully supported by vendors. To our surprise, we found this was not the case.

We found the current state of standards for languages and by extension development tools to be chaotic. While ANSI Standard C has indeed been with us for some time, we still have numerous compilers in the market which do not "quite comply". The new Object Oriented versions of C such as C + + have no agreed standard at all. We are seeing new and powerful language features which software engineers will insist on using in order to develop the ever more complex solutions demanded by users. Each use of these nonstandard features has the potential to "lock" in the applications to a specific platform supported by that specific compiler. What appears to be happening is technology within development languages and tools is moving much faster than the groups working on standards. Language standards therefore is an unresolved problem.

Next we looked at relational databases with the much referenced ANSI Standard for data base access, the SQL standard. This is frequently pointed to by hardware and software vendors as proof that a standard can be accepted, proof indeed that the commitment to Open Systems as represented by a willingness to accept standards is real. However, SQL itself represents a small set of instructions which in themselves are not complete or powerful enough to support the data manipulation tasks common within today's more sophisticated applications. DBMS vendors have written their own extensions to SQL which do provide support for a fuller functional data access mechanism but which if used make porting software to run on top of other relational data bases anything but transparent. Here again then, what on initial viewing looks like a problem resolved on closer inspection shows itself to be a problem blurred.

Another area of interest to the software vendor and ultimately to the end user is standards for the user interface. This has to be one of the most difficult areas to standardize as the user interface is a major tool in selling software. The ease of use, the appearance of the software, the methods for modification of data and screens, etc, is frequently seen as the "sex and sizzle" of the package. Functionality may be roughly equivalent across a number of packages but if the user can be persuaded that one software package is more "user friendly", that package will be an easier sell.

Today, as users frequently bring multiple third party packages into their environments, the confusing and widely differing user interfaces make it more difficult for the end user to move from one package to another. Application vendors who are attempting to write software which is portable across any number of hardware platforms have to contend with an almost endless number of graphical user interfaces which are promoted by the hardware vendors as unique to their environments. While the X-OPEN Portability Guide (XPG-3) does define standards for user interfaces which make porting much easier, there is little excitement among the vendors for restricting themselves to these standards.

The X Window system has become a defacto graphical user interface standard for UNIX systems. X Window is highly portable and is not proprietary. Working with something like the Motif window manager, it gives the user a very powerful and rich environment to work in. The problem is there are as many graphical interface products as there are vendors. Open Desktop is an example of one. It is a UNIX based graphical environment using the Open Systems Foundation OSF/Motif interface. It runs in a single or multi-user environment; the claim is that "its behaviour is compatible with Microsoft's Presentation Manager". There is some momentum building up for MOTIF to become the standard windowing system for UNIX.

Does this mean the graphical interface decision is an easy one? Not really; every vendor is developing his own interface all of which are unique, each of which has something exciting in itself for the end user. Nothing standard here for the poor applications vendor to latch on to.

Only in the area of communication standards is there much to praise. This may be for historical reasons but it is surely also because standards in communications continue to allow vendors to have some autonomy, some ability to differentiate themselves within their own domains. If Vendor A can communicate to Vendor B's kit without compromising anything he perceives as giving him a competitive advantage, then Vendor A will have nothing to lose by adhering to these standards. Indeed, adherence may allow him to put kit into sites previously closed to him because a different vendor's kit was already in place.

Communication standards, which for ease of discussion shall be referenced as OSI, are real. Products are being developed by major vendors which conform to these standards. Major exhibitions such as ENE88 demonstrate the practicality of linking multiple hardware platforms together into a single communications network. The major factor slowing the take-up of OSI among end users is cost which remains high reflecting the complexity of the solution. It is worth noting that this was also true with the initial introduction of Ethernet technology; over time the cost of Ethernet boards has come down to negligible levels as they became widely available through vendors and third party suppliers. The existence of other "standard" products such as TCP/IP and SUN's NFS which are now widely available from multiple sources reflect a history of acceptance of vendor developed technology across the wider community. If the same process occurs with OSI technology, then the take-up of OSI within the SME community will be facilitated.

What is different about communications is this:

Standards in this area are seen as a win for everybody. Vendors see communication standards as "opening up" previously closed areas, giving them opportunities to put replacement or upgrade systems into companies with existing IT capability. Applications providers see communications standards making possible such things as true distributed processing, EDI support, remote data entry and much more at a cost their user communities are willing to pay.

Larger users see OSI as protecting their capital investments. If one buys into communications standards, then in theory, kit purchased today will be compatible with whatever as yet unknown technology comes into the market tomorrow. Here then is benefit for the end user, the vendor, and the applications provider.

It is perhaps worth looking at why the SME user community has not yet accepted OSI as relevant to them. First and foremost, I believe the solutions currently available may not be especially relevant to SMEs. You may argue that connectivity is a common problem; that larger users simply have the problem on a larger scale. This may be logically correct but it does not take into account great differences in users perceptions of their own needs.

KEWILL has a large SME user base; we have discussed OSI as a solution with these users both in isolation and in user group meetings. We have found our users take the following view:

1.
OSI may be a solution but it is not one tailored to their problems.

2.
They have not been convinced that the solutions provided under OSI are cost effective.

3.
The initial capital investment is very high, enough in itself to put these users off. While larger concerns may feel they are spending a high percentage of their time and money integrating new kit, most SMEs feel they do not have that constant integration problem to contend with.

4.
The technology is not widely supported. If one invests in OSI solutions, it means committing to one of a small number of vendors who are offering OSI solutions. This lack of broad availability also means prices remain high.

5.
The technology may require support staff or staff training. So called proprietary solutions especially Ethernet solutions, are well understood by a high percentage

94

of engineers, etc.

6.
Finally, the solutions available today to SMEs work, they are cost effective, they are well understood. Our users are not sure they have the problem that OSI declares it solves.

A number of initiatives would be useful in increasing SME interest in OSI:

1.
An Awareness Initiative. SMEs must be better educated about OSI and in particular how it can benefit them specifically.

2.
A Vendor Initiative. SMEs are not the only ignorant members of the population. The vendors developing or marketing OSI solutions should be made aware of the profile of SME companies. They need to see that SMEs are not just like large companies, only smaller. SMEs typically are managed by a different breed of manager, they frequently run on much tighter margins, they usually are much less likely to make quick capital investments.

On the other hand, SMEs can be extremely innovative. They can move much faster than larger companies to embrace new techniques that work for them. There is much less bureaucracy to contend with in smaller companies. Learning about these variations can make all the difference in selling new technology to this user community.

3.
A Shop Floor Tools Initiative. OSI can only really succeed when one standardly buys new machine tools which are OSI conformant. Having an OSI network is only the beginning of the solution. We must also have a greater take-up of OSI among the manufacturers of machine tools. While we should expect currently installed kit not to conform to OSI, it is a sad comment on the industry that many new products announced donot include OSI conformant architecture. This means the installed base of non-conformant equipment grows every day.

3. Conclusions

The most disturbing thing about attempting to reach any conclusions about standards is how very many of them there are. The fact that so many standards exist and so many "draft standards are in review" makes them a source of amusement to outside observers. The state of Open Systems and the credibility of standards as a whole was recently surveyed by Price Waterhouse. They asked a number of IT managers who comprise the Price Waterhouse Opinion Panel what they thought about Open Systems and found that the majority of them did not believe the major vendors would ever be able to agree on standards.

Half the managers surveyed did not believe that POSIX would allow software to be readily ported to different vendor platforms before the end of the century. The majority of the managers did not see vendors moving towards resolving their differences in data base standards. In this market in particular, IT managers see vendors continuing to move apart, positioning themselves as optimized for UNIX or tailored for the VAX product line, etc.

The IT managers saw some hope for communication standards. Most of them expected standard interfaces to be a reality; 25 percent of them said such standards would be accepted within five years. Half of the IT managers expected to incorporate OSI within their IT strategies in the 90s.

Kewill, as a provider of CIM application software, is interested in developing software which will support the greatest possible set of user requirements, run on the largest possible range of platforms, and allow us to address the widest market sector possible. We want to ensure we incorporate major new technology into our product offerings and we want to bring the benefits of that technology to our end users. The only way it seems possible to achieve those goals is to design products which adhere to the evolving national and international standards, indeed to support OPEN SYSTEMS for SMEs. These standards may never be completely accepted, we may never have a definitive single standard for all, but surely, it is the only logical starting point today.

It is interesting to speculate on what could be done to motivate more IT vendors to build products to the major standards discussed within this paper. The motivation can only come from the market; when SMEs begin to expect their software to conform to Open Systems standards, when they include these standards in their Invitations to Tender (ITTs) or in their basic requirments lists, then vendors will respond by offering conformant software.

Kewill believes that software built to these major standards not only brings the benefits of those standards to the end user, it offers us a competitive advantage. As long as some IT vendors market products to SMEs which are not conformant, the more innovative and aware vendors such as Kewill can benefit by offering "insured solutions"; Solutions which are insured against the inevitable technical innovations in the next issue of any IT magazine.

Kewill is confident of the correctness of this approach. In conjunction with British Telecom and The Dale Group, a holding company with multiple operating sites with SME profiles, Kewill has just initiated a project to put OSI solutions in place in five user sites within the Dale Group. We will be developing a case study as part of this project which will set objectives and measure the solutions provided against those objectives. In this way, we will at last have a real world cost benefit analysis backed up by quantitative data which show the benefits SMEs can realize from OSI.

The result of this project should be an increased awareness of the benefits of OSI in the manufacturing market and the ability to quantify that benefit within the SME community.

4. Acknowledgements

Kewill has received a number of matching grants from the Department of Trade and Industry (DTI) who share our interest in bringing the benefits of Open Systems to SMEs. We acknowledge our appreciation of their financial support but are even more grateful for their continued interest and belief in our ability to develop meaningful solutions for our users.

I am also grateful to British Telecom and in particular to Norman Sigrist and David Gallagher who persisted with me in a long search for an appropriate SME end user site for this project. BT's expertise in OSI and their interest in OSI as it is related for solutions for manufacturers has lead to a collaboration with Kewill which will result in tangible products supporting standards and bringing improved communication platforms to SMEs.

Finally, I am grateful to Chris Coole and Tony Graves of Dale who share our vision of the future and are willing to go through prototype hell to get there.

5. References

(1) IEEE Standard, 1003.1, published by The Institute of Electrical and Electronics Engineers, Inc.

(2) Information Technology Review, Price Waterhouse

* Footnote: UNIX is a Trademark of AT&T.

Implementing CIM : A Vendor Perspective

N.E.Brownlow

Groupe Bull
Manufacturing Marketing and Business Group
1, rue Carpeaux
Cedex 74
92039 Paris La Défense
FRANCE
Tel:- +33 1 46 96 85 47

Abstract

The implementation of Computer Integrated Manufacturing (CIM) has become the subject of great attention in recent years. In particular, technologies such as local area networks have gained prominence due to their fundamental importance in any CIM architecture. Within ESPRIT several initiatives are taking place to progress the technology such as the CNMA, DELTA 4 and FICIM projects, however, the uptake of the existing technology remains slow.

In this paper the author takes a look at the implementation of these technologies from a vendor point of view and in particular looks at Manufacturing Automation Protocol (MAP) and its position in the CIM environment. Some insights into the slow uptake of the technology are offered and the benefits which might be gained by those who started to adopt the technology explained.

Summary

In this paper the importance of communication technology in achieving competitive advantage is presented. The benefits for those who have started early in implementing the technology are examined and the areas where work still remains to be done are highlighted.

An insight into where ESPRIT CIM programs are addressing the outstanding needs is given and in particular the goals of the Communications Network for Manufacturing Applications (CNMA) project and its relationship to MAP are examined.

1. Introduction

Manufacturing Industry is constantly seeking ways of improving efficiency and effectiveness by the adoption of new technology and working. Two philosophies, in particular, stand out from the rest, those being Just in Time (JIT) and CIM. Both are widely offered as a panacea for the problems facing manufacturing industry, however it is now realised that you cannot buy either of these technologies "off the shelf".

There is no single CIM system that can satisfy the needs of all companies and the implementation of Just in Time never ends, rather it is an on-going quest for the elimination of waste within the enterprise. However if one were to take the principles of these two philosophies and apply them to a particular factory or problem significant benefits can result.

The question is how does one go about this task with the currently available technology.

2. The Strategic Importance of Communication

For centuries man has realised that timely, accurate information is a significant weapon against competition. The use of watch towers with communication by semaphore warned defending armies about the movement of the enemy and has today been replaced by modern communications technology in the field. In civilian life the telephone has become commonplace in most civilised countries and the ability to dial anywhere in the world taken for granted. The ability to communicate with stockbrokers, news services, suppliers and customers has been used by many companies to gain competitive advantage in the market place and to create new market opportunities. The mail order business is one such example giving the customer the ability to purchase from his/her own home via the telephone and making payment by credit card.

As in the case of the telephone system which grew from a number of isolated local exchanges, to interconnected national networks and finally to international telecommunications systems, the need for internationally recognized standards has been recognized in order to interconnect the various components of the system. Today the "standard" telephone is treated like any other household item and not many companies in the 1990's could exist without a telephone connection. In Alexander Graham Bell's time the answer to this question was probably not so obvious.

The manufacturing market place is becoming more international and therefore more competitive. Companies that have in the past enjoyed near monopolies in their own country have suddenly found themselves in competition with cheaper imported products. This trend has exposed the inflexibility of many manufacturing companies as they have found themselves unable to change their working methods in order to respond to the needs of the market.

In the 1980's quality was the goal of most manufacturing enterprises, today quality is taken for granted and the flexibility to respond to customers specification and timing requirements are now just as important.

These requirements have exposed the often lengthy lead times necessary to get a product from design to manufacture and short term solutions such as increasing stock levels have highlighted the cost of holding such stock.

Here lies part of the interest in the afore mentioned philosophies.

JIT looks to eliminate waste at all stages of the process, such as excess stock, work in Progress (WIP), scrap, etc.

CIM offers the ability to decrease the time taken to get a product manufactured by for example being able to use information produced by Computer Aided Design (CAD) systems to produce the necessary part programs for CNC machines and transfer the necessary Bill of Materials (BOM) data to Manufacturing Requirements Planning (MRP).

The effort required to obtain simple information as background for JIT improvement programs has highlighted the inadequacy of some of the information systems currently installed and the requirement to mix information from different sources has exposed the incompatibility of the various computers, networks and databases.

Clearly, those companies who can supply timely and accurate information to their decision makers in the desired format stand to gain in the market place provided of course the information is acted upon in an appropriate manner.

3. The Use of Standards

As in the example of the telephone system the ability to link the various components of the communications infrastructure together is simplified by the adherence to standards. The term "OPEN SYSTEMS" is now widely used to refer to systems that conform to international standards however this term is subject to some abuse.

Many companies have merely published their own proprietary standards and placed them in the public domain declaring them "Open". This offers no benefit to the user unless all major companies support the standard in which case it may become a de facto standard such as say the personal computer compatible.

The real benefit to users comes from the implementation and use of internationally recognized standards who's control is preferably in the public domain.

When it comes to communicating with shop floor devices the integration of such devices to other systems can become a nightmare of incompatible physical media followed by incompatible communication protocols, a problem that faced General Motors when they created the Manufacturing Automation Protocol (MAP). MAP offers a standard way of connecting manufacturing devices together in order to achieve an integrated system.

The political controversy that surrounds MAP is not covered in this paper however despite this controversy there still remains no other offer on the market place capable of meeting the true "Open System" needs.

MAP's benefit to the user is often quoted but from a vendor perspective MAP also offers similar benefits. Currently vendors such as ourselves have to support a variety of proprietary protocols in order to allow our customers to connect their computer systems to our manufacturing devices. The development of these communications links are expensive and these costs are eventually met by the customer. MAP offers the opportunity to reduce this burden. Another advantage for vendors lies in the communications media access for applications programs. MAP offers a standard interface in the form of Manufacturing Message Specification (MMS) which allows communication between the application and the shop floor data.

It would be naive to think that all vendor companies would support MAP at the expense of their own proprietary systems. We ourselves are in a fortunate position of having no proprietary system to protect and therefore welcome the MAP standard however other vendors have several millions of pounds invested in proprietary solutions and wish to protect their installed base.

This last fact is part of the reason for some of the controversy over MAP and the lack of availability of MAP devices in the past. Another reason for this controversy was the lack of a migration path from MAP 2.X to MAP 3.0 and the announcement of the MAP 3.0 specification just as MAP 2.X product was about to hit the market. As a result of this a six year freeze was placed on the specification in order to allow implementations to take place. It was assumed by many that this meant that the MAP 3.0 specification would not change during this time period however the statement really only assured upward compatibility of future MAP versions such as the MAP 3.0 (1991) version. To a vendor this seems a somewhat strange state of affairs since one would assume a user driven standard would ensure upwards compatibility between all versions without any time limit. Certainly a vendor could not afford to abandon a customer after six years just because the MAP compatibility statement had expired.

MAP's intention is to use international standards as they become available and therefore will need to change its specification as those standards evolve. There are clear signs in the market place that ISO networks will take a major slice of the communications business and the future of these types of network is assured.

Today one would never think of installing a telephone system that was incompatible with the standards of the national telecommunication network, those who are planning for tomorrow will not make a similar mistake when planning their computer communications.

4. MAP - CNMA Relationships

MAP although a major step forward is by no means complete in terms of offering a solution to computer communications problems. In terms of international standardisation several of the protocols referenced in the 1988 version of the specification have moved on in the international standardisation process, notably MMS and Network Management (CMIS/CMIP). These International Standard (IS) versions of the protocols are not entirely compatible with the MAP versions and thus MAP is somewhat tied in having to the keep the old versions of the standard in order to honour its upward compatibility statement.

The CNMA consortium members have agreed to act as "consenting adults" when it comes to the use of emerging international standards and has endeavoured to track these standards as they have progressed through the various standardisation bodies. In particular MMS and Network Management have been areas where the work done in the consortium will have future benefit to the MAP community. Thus CNMA is in advance of MAP, however, all of the CNMA work has used the MAP standard as its base and compatibility between CNMA and MAP devices can be demonstrated.

CNMA has also proposed certain extensions to the MAP specification, probably the most notable being MAP over Ethernet and will continue to work in this direction based on the input from the CNMA users.

The answer to the question is CNMA a rival alternative to MAP is NO. Both have a common goal, the use of international standards to realise a communications network for manufacturing applications and should therefore be able to reach a common conclusion in the future. Indeed, there exists a real opportunity for closer collaboration with those planning the future of MAP and those pioneering the CNMA technology. MAP needs to ensure that future specifications are unambiguous and implementable and can therefore benefit from CNMA experience. CNMA can ensure that its work is taken into account in the future MAP specifications and therefore in the market place.

Some examples of where this cooperation could be achieved are in the areas of MAP over FDDI media and Remote Database Access (RDA) both subjects being addressed in the next CNMA phase. In the case of remote database access the goal is to achieve transparent access to heterogeneous databases distributed over an ISO network. This work is fundamental to the realisation of true CIM. The advantage of such a co-operation for users is the benefit of a window through which they can see the future direction of the MAP protocol, something that is very difficult to achieve at the present time despite the efforts of the user groups.

5. Implementing the Technology

Unfortunately the technology of CIM is sometimes seen as a panacea for the problems of manufacturing industry and in cases where this has happened it is often the technology that is blamed for the failure. The old adage "Garbage in,

Garbage out" is equally true of computer communication as it is of the computer itself. It should be obvious that if a system is so complicated that people are overloaded with paperwork the installation of a computer will merely generate the paper more quickly adding additional pressure to the already overloaded system.

The lesson of JIT is keep it simple, if it is necessary do it, if is not don't. Simple systems are easily followed and understood and because they require only the bare minimum from the user they are often respected and used. After arriving at this stage the next step is to integrate these simple systems into a whole making minor adjustments to the system in order to cope with the added complexity.

Finally, if this stage is reached successfully, computerise and automate those areas of the system where an increase of the overall efficiency can be realised.

This final point implies that not all of the factory will be computerised and indeed this is the case. The so called fully automated factory will be the exception rather than the rule well into the next century. The requirement to take note of the views of people within the CIM environment cannot be overlooked as they form a vital link in any communication system.

6. The Future

If we accept that the use of Local Area Networks (LAN's) based on MAP are to become the heart of the communications system on which the company depends for its very survival then it becomes obvious that the failure of such a network would have disastrous consequences for the company. The use of fault tolerant techniques to protect the functioning of individual computers in the presence of failures is fairly well accepted however, protection against failure of the communications network is less widespread. Within ESPRIT the Delta 4 project attempts to deal with both the issues of network and computer system fault tolerance by the use of redundant media for the communications network and also by using the network itself to guarantee fault tolerance of the computer applications.

The basis if this work has been the CNMA communications stack and Delta 4 maintains compatibility with CNMA and MAP devices. A pilot facility will be installed in a Renault car plant during the course of the project.

Another important development taking place is the formation of an internationally recognized standard for a fieldbus. Fieldbusses are relatively low cost networks, based mainly on twisted pair cable, used for communication with sensors and actuators on the shop floor. It is apparent that the cost of MAP will prohibit its use for such devices and hence the requirement for a fieldbus.

The fieldbus architecture is designed with high performance in mind and the question of whether there is room for both the higher performance, reduced functionality MAP specification, Minimap, as well as a fieldbus remains to be answered.

Within the fieldbus arena two standards are prominent in Europe today, FIP and Profibus. The ESPRIT project FICIM aims to integrate these two fieldbusses into the CIM architecture by developing a connection to CNMA networks. This project will prepare the way towards the implementation of an international standard solution in the following phases of the project.

Whatever the final versions of these technologies it is certain that they will need to be fully integrated into an ISO "Open Systems" environment.

7. Why Invest in MAP Today

Some companies are playing a waiting game with MAP to see if the technology becomes cheaper and more complete. Whilst this will inevitably happen resulting in short term savings for these companies they may be at a disadvantage in the long term.
Those companies implementing the technology today are in the learning stage both in terms of how the technology works and how to work with the technology and as a result are increasing their skills base. However in implementing the technology they are putting in place the necessary infrastructure in terms of the backbone networks, MAP platforms and MAP applications which will allow them to take early advantage of the enhanced versions of the technology. Companies who have not taken these steps may well be faced with the problem of having to invest heavily in the future in order to remain competitive and acquire the necessary skills and communications infrastructure.

In many cases it is unlikely that the necessary investment can be made available over such a short time period and that, at best, a number of years will be required in order to regain a competitive position.

Finally some companies have already made short term benefits through the implementation of MAP carrierband technology which is becoming widely regarded, by those who have used it, as being cost effective when compared to non MAP solutions.

The message is that a long term strategic approach is necessary when dealing with communications technology. However, the building blocks necessary to realise the future strategy are available today and if used correctly will ensure the future competitiveness of the company.

Conclusion

There is some way to go before fully integrated information systems become a reality however the rate of progress is increasing. The building blocks are available today in the form of MAP standards and allow users to familiarise themselves with the technology. Those that take this opportunity will benefit in the long term from being able to capitalise on the future technology. ESPRIT programs such as CNMA, Delta 4 and FICIM are contributing to ensure that a complete solution to the problems of manufacturing industry can be realised in a relatively short time and that these programs are following the standards adhered to in todays products.

References

[1] Manufacturing Automation Protocol V3.0, August 1988

INTRODUCTION TO CNMA AND MAP

COMMUNICATIONS FOR MANUFACTURING
A REVIEW OF THE CNMA PROJECT

Tim Simmons

BAeCAM
British Aerospace
The Guild Centre
Preston

Summary

The communications Networks for Manufacturing Applications project is specifying, implementing, validating, and promoting emerging communications standards for CIM. This highly successful project has commissioned a number of real production pilot facilities and provided working demonstrations at three major international exhibitions.

1.0 Introduction

If European industry is to be competitive in world markets, it must make good use of Computer Integrated Manufacturing, CIM. For optimum profitability, a manufacturing facility must be flexible and reliable. It must be able to produce quality. It must provide accurate, timely feedback and must respond quickly to management decisions or new requirements. To be competitive, European industry must ensure that manufacturing resources are fully integrated.

A wide range of sophistocated computer based devices is now available for use in or around the shop floor. Most of these use different proprietary languages to communicate. Integration of these devices is expensive and slow because bespoke interfaces are required. What is needed is a standard communication language made available for everyone to use. The development of such a language is a complex, expensive task. To be successful the standard language must fulfil the needs of a wide range of users, and it must be adopted by a wide range of vendors. The task of specifying the language therefore requires input from, and collaboration between, many different companies and organisations.

The Commission of the European Communities has provided a framework and funding for such collaboration, through the European Strategic Programme for Research and Development into Information Technology, ESPRIT. One of many projects supported in this way is CNMA, which addresses Communications Network for Manufacturing Applications.

CNMA's objective is to undertake pre-competitive, collabarative activities to stimulate the growth of the European OSI communications market, and to enhance Europe's competitive position in that market. CNMA's direct effects are :-

o   Development of European expertise in OSI communications

o   Assisting and influencing the development of communications standards

o   Encouraging the production of OSI conformant products and test tools

o   Developing strategies for migration from current proprietary solutions to OSI

2.0 Procedure

The CNMA project is now in phase 4. During the course of the previous phases of the project a well defined procedure has been developed. Taskswithin the project fall into seven major categories. These are User Requirements Analysis, Specification, Implementation, Testing, Pilot Demonstrations, Information Dissemination, and Standards Activities.

2.1 User requirements analysis

This activity is designed to ensure that the project as a whole keeps in touch with the real requirements of users. The User partners come together to produce a User Requirements Report. This is a long range analysis, in some cases defining CIM problems which we do not expect to be solved by standards or products for as long as five years in the future.

2.2 Specification

The first task for the Vendor partners is to agree on the communication profile which is to be used in the particular phase of the project. This is a very complex task requiring a careful balance between on the one side the desire to be compatible with developing standards, and other developing agreements world wide, and on the other side the need to agree a complete profile which contains all details necessary to allow the profile to be implemented. This activity leads to a clear unambiguous specification of the communications profile to be used, which is then set down in a document called the Implementation Guide, (IG).

2.3 Implementation

Once the Vendors Partners have agreed the communication profile, the Vendors can then complete the task of producing communication software for their own equipment, that conforms to the profile. This is largely a separate activity, with each Vendor company working individually, and using the Implementation Guide (IG) as the specification for its own software. This has interesting effects. In general the same people are involved in Specification and Implementation. The fact that they all know they will have to produce communication software solely from the IG ensures that the IG is of good quality with all potential conflicts and confusion sorted out at the specification stage. Furthermore, because a number of separate groups of people around Europe are each responsible for producing software implementations, each group is well motivated to ensure that it thoroughly understands and becomes experienced with the technology - thus ensuring that European experience is efficiently developed.

2.4 Testing

One of the early lessons learnt within CNMA, and domonstrated now many times, is the value of formal testing software. This ensures that each implementation of the CNMA profile really is conformant to the IG.

Before any communication implementation is used in any application it is rigorously tested by these test tools, produced by the associated ESPRIT project TT - CNMA. Once this is done, User partners can be confident that all communication software is of good quality, and is reliable, which in turn greatly eases the commisioning and installation activities in later tasks of the project.

2.5 Pilot demonstrations

In each phase of the project the CNMA profile is deomonstrated in a number of Pilot applications. Some of these pilots are real manufacturing facilities, using equipment from at least two CNMA Vendors. Other Pilots are experimental facilities, combining equipment from all Vendor partners. Other Pilots may be dedicated demonstration facilities, or test beds for particular functions of the CNMA profile. In practice the Pilots within a particular phase represent the requirement which must be satisfied by the CNMA communications profile and therefore have some influence on the communication services which are adopted for that phase. It is therefore important that pilots are selected to represent a broad range of common communication requirements. This needs a synthesis of the requirement specified by Users Requirement Reports, and the requirements of Users as perceived by Vendors through their normal market awareness activities.

2.6 Information dissemination

This is a continuous responsibility for CNMA. The project publicises the results of its work to stimulate interest in OSI communications. This is through events such as this "Communication for Manufacturing" conference, through presentation of papers, press releases, magazine articles. Indeed in some cases, attendance at exhibitions has meant providing major OSI demonstrations as was done at Hannover in April 1987 and at ENE88i at Baltimore in June 1988. This has also been arranged for this Communication in Manufacturing conference. Visits are being laid on to view the first Pilot demonstrator of the current phase of the project which has been installed at the University of Stuttgart.

2.7 Standards activities

CNMA has to maintain continuous liaisons with standards bodies world-wide to ensure that we collaborate rather than compete. In general, the liaisons are at a detailed technical level, but may have strategic effects on the CNMA project. To control this target objectives have to be defined, which can be matched to the CNMA technical strategy.

3.0 CNMA EP 955

The success of this procedure was demonstrated in previous phases of the project by a number of Pilots.

3.1 Hannover Fair Demonstration - 1987

This highly successful CNMA demonstration was the first public exhibition of computers and controllers from different vendors interworking to control a manufacturing cell, using CNMA communications software. As such, this demonstration was an important milestone in the development of standards to allow full interoperability of computers and in the advance towards truly Computer Integrated Manufacturing.

The advanced manufacturing cell incorporated an Automax Machining Centre, driven by the CNC, and a transporter and a robot, each controlled by a PLC. The cell control, including operator-access, was provided by the mini-computers and each of the relays linked two of the Local Area Networks. TITN, the systems engineers, wrote the cell control application software.

3.2 BMW Production Demonstrator

February 1988 saw the commissioning of CNMA's first production demonstrator at BMW's new car factory in Regensburg, Bavaria, West Germany. The CNMA communications system looks after the 'Real Time' transmission of production line error information which is vital to the maintenance of the 'Just in Time' concept of the plant, and this application underlines BMW's confidence in the work of CNMA project.

The production system, which is the World's first factory floor implementation of the manufacturing Message Specification uses a Siemens mini-computer to collect error information generated by up to one hundred PLC's controlling the production line, and any critical error messages are transmitted immediately via a GEC router to the Nixdorf mini-computer as part of a comprehensive plant level maintenance system.

3.3 Aeritalia Production Demonstrator

This demonstrator was commissioned in October 1988, and controls the cutting and stamping of military aircraft electrical wire harnesses. The Turin-based facility was linked to the British Aerospace facility providing a demonstation of a mnaufacturer's ability to transmit production data to sub-contractors, quickly and reliably.

3.4 British Aerospace Production and Enterprise Networking Event
    Demonstrator

In January 1988, testing started at British Aerospace Salmesbury, the site of CNMA's second major production demonstrator. CNMA computers, controllers and communications software controlled a flexible manufacturing system used for the production of A320 Airbus components. This facility formed the basis of CNMA's participation in the Enterprise Networking Event, held at Baltimore, USA, in June 1988, and was the only real production facility participated in the Event. CNMA devices communicated, using FTAM, MMS and MHS with devices at Baltimore, providing an important networking demonstration.

4.0 CNMA EP2617

The current phase of the project is called CNMA EP2617.To provide the correct combination of skills, interests and resources, CNMA EP2617 comprises a consortium of 17 European organisations.

Users	British Aerospace (Co-ordinator)
	Aeritalia
	Aerospatiale
	Magneti Marelli
	Renault

Vendors	Bull
	GEC
	Nixdorf
	Olivetti Information Services Ricerca
	Robotiker
	Siemens

Systems Engineers	Comconsult
	Fraunhofer (IITB)
	Alcatel TITN
	Syntax Sistemi Software

| Academic Institutes | University of Porto |
| | University of Stuttgart (ISW) |

CNMA EP2617 is building on the results of CNMA EP955. Work is concentrating on the following areas :-

o   The communication implementations produced in CNMA EP955 are being updated to reflect the advances made in the international Standards bodies.

o   Directory Services is being specified and implemented.

o   Standardised application interfaces for FTAM and MMS are being developed to assist application portability.

o   Network Manangement Functionality is being enhanced by the addition of new services and by increasing the number of managed attributes, particularly in the upper layers.

o   An automated Network Management application is being developed which uses knowledge based techniques. This application will for example be used to automatically optimise performance, configure the network, and diagnose fault conditions.

o   Four pilot applications are being commissioned. These are at the University of Stuttgart, Aerospatiale in Chatillon, Magneti Marelli in San Salvo and Renault in Boulogne Billancourt.

4.1 ISW Experimental Demonstrator

The demonstration facility at the University of Stuttgart (ISW) consists of two cell. One cell features a turning centre, a boring and milling centre, a linear portal robot, and a pallet store. Production will be fully automated. The other cell has manually loaded 5-axis milling machine, demonstrating the advantages of Local Area Network Communication betwen Computer Aided Design and Numerical Control Systems.

Thirteen processors from six European vendors will be linked via three local networks providing Computer Aided Design and Planning, Part Programme Management, Area, Cell and machine control functions. The pilot facility at ISW demonstrates the latest OSI communication services, and the migration path from proprietary protocols to an OSI environment. A Nixdorf Targon 35 supports Computer Aided Design and order entry functions, whilst an Olivetti LSX 3020 computer provides a file server for management of part programmes.

Control of the Linear Portal robot is by a GEC GEM 80 PLC. The pallet store is controlled by a SIEMENS S5 SIMATIC PLC.

A gateway device from ROBOTIKER supports communication between CNMA protocols and the five-axis NC machine. SIEMENS SINUMERIK devices provide numerical control of the machining and turning centres. These devices use the proprietary SIEMENS SINEC H1. A Siemens Gateway device is used to provide communications between the proprietary and OSI environments, demonstrating a migration path. Systems integration is provided by ALCATEL-TITN in conjunction with ISW.

The software is run over three Local Area Networks, an 802.3 CSMA-CD Baseband LAN as specified by TOP, and 802.4 Broadband LAN as specified by MAP, and an 802.4 Carrierband LAN for low cost MAP implementation. These three networks are linked by router devices from GEC and Bull.

In addition to the MMS and FTAM services used to integrate the applications, a major implementation of Network Administration, is supported by all the vendors. A Fault and Configuration Management System is provided By BULL and the Fraunhofer institute, whilst a Performance Management System is provided by SIEMENS.

4.2 Aerospatiale Demonstrator

At Aerospatiale in Chatillon, France, nine machines will be controlled in the manufacture of prototype missile components. The main objectives of the project are to improve on delivery times, quality and flexibility. A Maintenance Management Application will collect status information from the machines and produce fault diagnosis by expert system. A shop Management Application will perform short term planning and will produce machine shop performance statistics. A Machining Cell Management Application will take job schedules from the Shop Management Application and will handle all data communications to and from the NC controllers. Finally a Storage, Handling and Transport Cell Application will control the AGV transporter, the manual load/unload stations and the storage area.

4.3 Magneti Marelli Demonstrator

This pilot is at Magneti Marelli's factory in San Salvo, Italy. It will be installed in the first section of an alternator production line. Three main functions will be performed. A monitoring system will collect data on machine productivity and behavior, a tracking system will trace all work in progress on the shop, and a diagnostic system will collect detailed information on machine states. The Demonstrator includes both OSI and proprietary devices, thereby demonstrating migration to OSI.

4.4 Renault Experimental Demonstrator

At Renault in Boulogne Billancourt, France, a test and demonstration facility will be set up for Network Adminstration Systems.

In large networks, automated facilities are required to handle re-configuring of the network.

Suitable protocols for this are provided by Network Manangement. The pilot will have a conference room linked to a test laboratory, where three different demonstrators are performed. One of these will be a simulation of the Aerospatiale pilot. The second will demonstrate control of video equipment from the conference room and the communication of video images via the network. All three demonstrations will be Lined Area Networks. In combination, this will illustrate the ability of Network Management to tune and reconfigure the network, and diagnose and manage fault conditions. Once the Network Management Application has been proven on the Renault Experimental Demonstrator, it will be installed in the Aerospatiale Demonstrator, thereby showing its value in a real production environment.

5.0 Summary

In summary, the major achievements of the project can be listed as follows :-

a)
Four Implementation Guides have been produced defining a communications profile to allow MMS, FTAM, Network Management and Directory Services protocols to be exchanged between multi-vendor systems of mini-computers and programmable devices.

b)
Multi-vendor control using CNMA communications software has been successfully demonstrated at the 1987 Hannover Fair, the Enterprise Networking Event, in a number of real production environments and is being demonstrated now at the ISW Pilot at the University of Stuttgart. A number of world firsts have also been achieved during these activities.

Some of these pilot demonstrators incorporate both proprietary and OSI communication, thereby demonstrating the ability to migrate from one to the other.

c)
CNMA is liaising with standards bodies and is having an impact on the standards, thanks to its experience in implementation and validation of the profile.

d)
CNMA vendors now have a number of OSI based products on the market, and a comprehensive set of MAP 3.0 test tools are now marketed world wide by SPAG-CCT, a commercial venture launched specifically for the marketing of the conformance test tools.

USER REQUIREMENTS FOR COMMUNICATIONS

Dr. J.B. COX

British Aerospace (Military Aircraft) Ltd.
Warton Aerodrome
Preston
Lancashire PR4 1AX
United Kingdom

Summary

The ESPRIT Project 2617: CNMA User Requirement Study (Task 202) being carried out by British Aerospace, Aeritalia and the University of Porto is on target for completion at the end of 1990. This note reviews the methods adopted for the study, the progress to date and the main requirements.

1.Introduction

The computer industry has evolved rapidly through a number of phases in the past twenty years. Initially "large" mainframes were used to automate repetitive, mostly clerical, tasks and their use was closely controlled by a Data Processing Department.

Technology advances then made minicomputers sufficiently economic to be used for specific applications in many small areas of a company. Control started to move away from the Data Processing Department to local Departments where the users were closely involved in the development and support of self-contained application packages. This moved further to the control of the users with the widespread introduction of personal computers and workstations with off the shelf software.

In the third stage of evolution the technology is being developed to link these separate application systems together to exchange data over networks. This development is hampered by the multiplicity of data formats and hardware connections. "Standards" have been developed as required to simplify the data exchange process in particular application areas by hardware vendors. The 1989 OMNICOM Index of Standards contains reference to over 2700 standards relating to networking and Data Exchange!

The next stage of evolution is to move beyond the concept of Data Transfer and Exchange to Information Sharing where Information Management is used by organisations for competitive advantage. This stage is characterised by being driven very much by users requiring The best software package on the most appropriate hardware for each application (or part of an application) with the expectation - no certainty - that it will integrate with the rest of their applications. It is the requirements of this new age that is addressed by the current study. With this stage of evolution we have arrived at a computing environment depicted in fig.1 in which the network is central to the Organisation and where the user may never need to know which computer (or computers) is performing the required operation. This computing environment is assumed throughout the study.

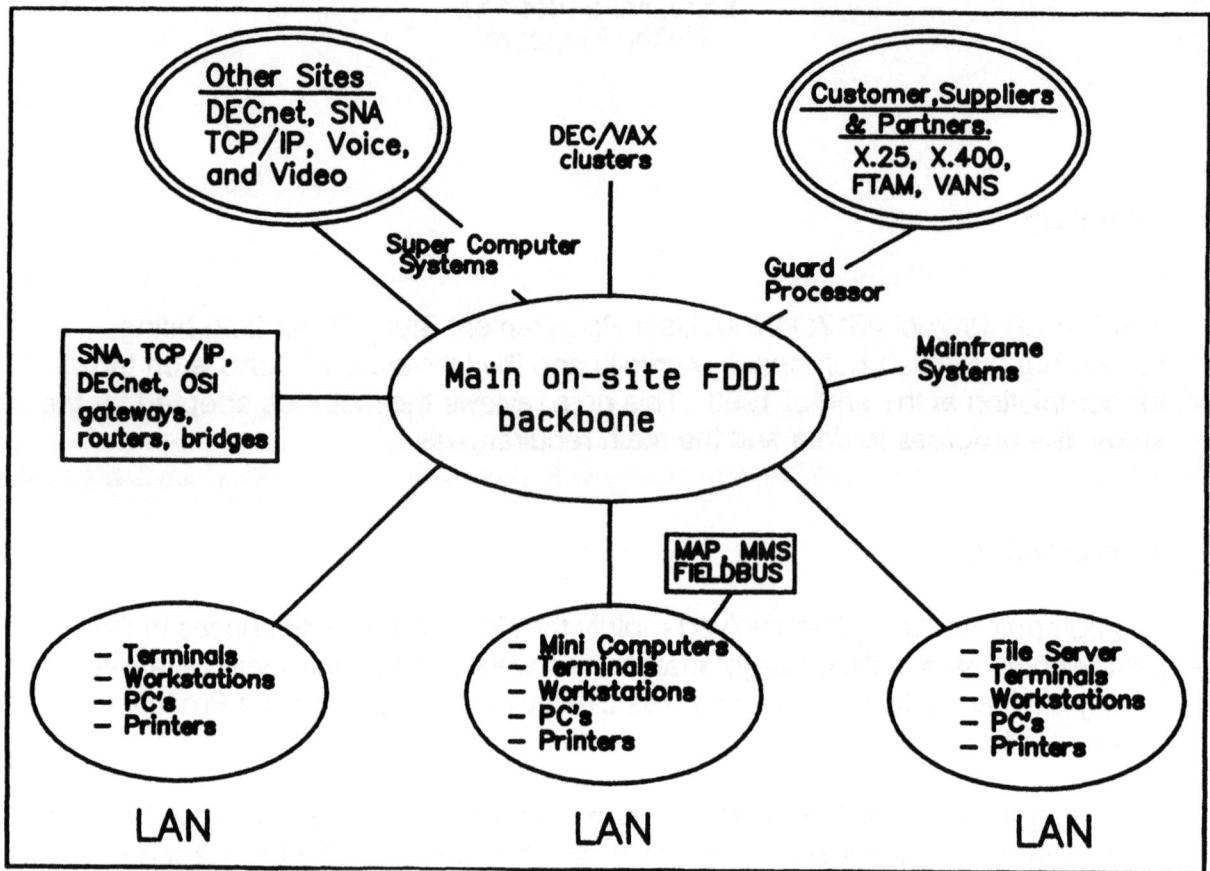

Fig.1   Projected Computing Environment

2.Importance of communications

Although the CNMA project is concerned primarily with Networks for Manufacturing Applications, the requirement for Information Sharing necessitates the view of information throughout the organisation. The importance of data to an organisation cannot be emphasised too highly. The ability to get quality products to market quickly has never been more important than in today's highly competitive global market-place and in order to sustain a competitive advantage those products must be continually upgraded in terms of reliability, maintainability, productibility and acceptability in the market-place. Manufacturers are embarking on strategic initiatives such as Concurrent Engineering, Total Quality Management and Design for Manufacture in order to reduce time to market.

Underpinning all these concepts is the necessity to have the right information available to whomever needs it, whenever it is needed, wherever it is needed and in the format required. The management of data has always been important whether held on computer file or on paper or microfiche. What is of paramount importance in the next stage of computing evolution is the ability to communicate that data speedily and accurately, that is to transfer and interpret the data.

This is the case for all manufacturing industry but is perhaps taken to the extreme in the Aerospace Industry where vast amounts of data are generated relative to the number of items delivered and where product life cycles are typically thirty or forty years from inception to the last aircraft in service.

3. Method of study

The current study of User Requirement is being carried out in two phases. The first phase considered an organisation split into a number of Business Activities and investigated the data communication requirements of each Business Activity. Business Activities considered were:-

- . Project Management
- . Technical Definition
- . Engineering Definition
- . Factory Scheduling and Control
- . Manufacturing Control
- . Commercial Systems

and two technology areas applicable across each Business area:-

- . Security/Data Management
- . Communications Utilities

Note that these Business Activities do not correspond exactly with the CIM-OSA Business Model (ref.1) but it is intended to reconcile the differences in the final report.
The relationships between the Business Activities (and within each activity) are complex. Information flows as shown in fig.2 were considered in terms of:-

- . type of data
- . quantity of data to be stored/transferred
- . frequency of transfer

Phase 1 of the Study was completed in 1989 and a "User Requirements Study Interim Report" has been supplied to the CEC as the first deliverable of the project.

In the second phase of the study the initial requirements are being compared against the availability of international standards and the requirements for migration from the current environment to where we think we wish to be are considered. In order to put some structure into the requirements, the report has been constructed around a Business Model known as the ABC (Architecture for Business Communications) which is a derivation of the work carried out by APEX (Advanced Project for European information eXchange). APEX was originally a EUREKA funded project which developed the APEX Communications

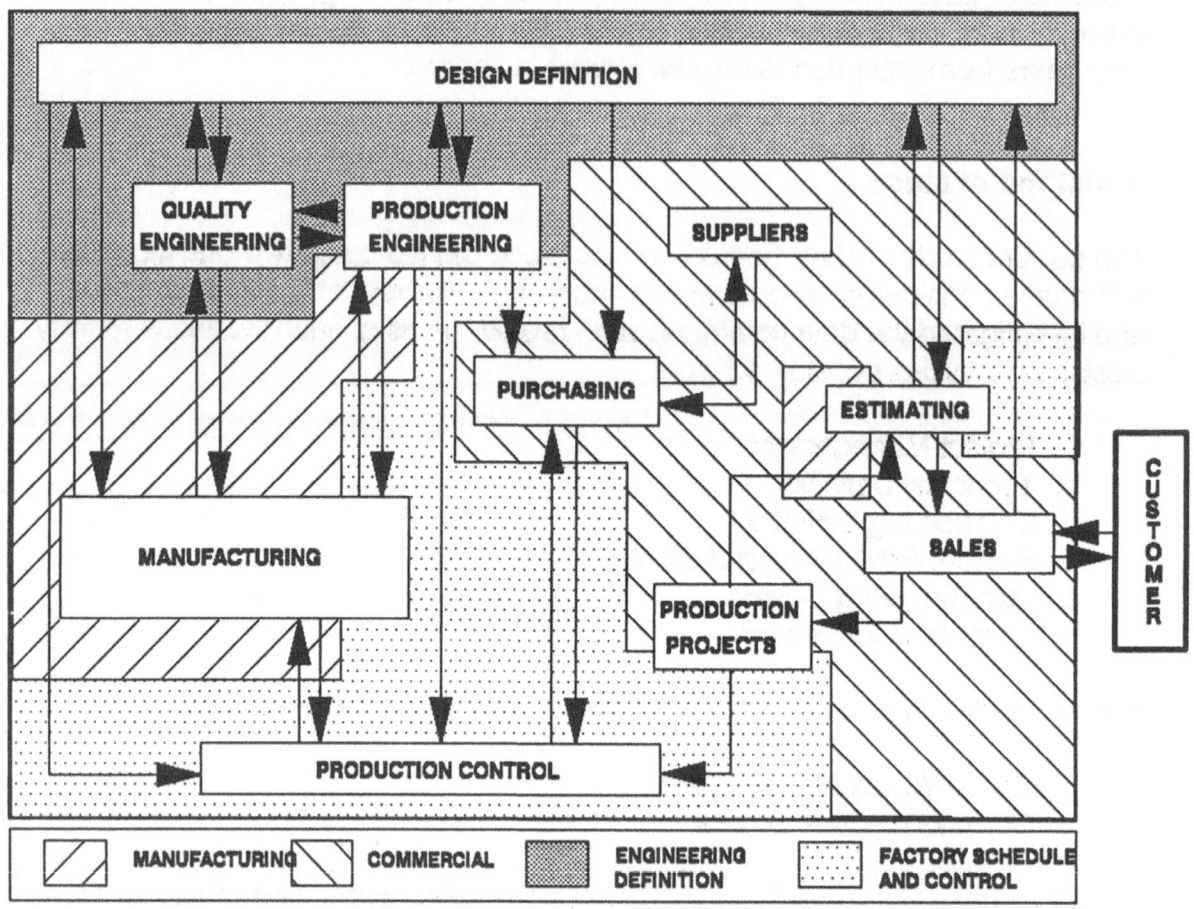

Figure 2. The Business Enterprise

and Network Architecture - abbreviated to ACA and ANA respectively (refs 2,3). APEX is now a separate Organisation sponsored by a consortium of European aerospace companies (Aeritalia, Aerospatiale, British Aerospace, Messerschmidt Bolkow Blohm and CASA). The ABC model was chosen in preference to the more conventional ISO 7 layer model as being more closely related to the User perception of application communications than the ISO model, which is more technically based. Within the User Requirement Study some of the layers in the APEX model have been expanded from their narrow communications viewpoint to cover a wider computing interpretation. The User Requirement Model (with deviations from the APEX model noted where appropriate) contains four layers which are surrounded by legal and commercial constraints. These are shown in comparison to the ISO 7 Layer Model in Fig.3.

The general User Requirements from Phase 1 are being reviewed against each of these model layers together with relevant standards and recommendations will be made.

Three projects currently being carried out within British Aerospace and Aeritalia are being used as pilot projects to evaluate the requirements in real situations and to identify shortcomings and problems. These projects cover:-

. Engineering Information Management
. Distribution of Engineering Information
. Electronic Data Interchange (EDI)

Phase 2 of the study also includes a review of one of the key requirements of the User community. This is how to develop a strategy for migration from today's environment to the one we are defining for the future, taking into account the minimum disruption to the business and cost constraints.

```
_____
|       ABC MODEL                                      ISO MODEL        |
|_____|
|  .   Information                                                      |
|                                                                       |
|  .   Data (APEX: Data Formatting Methods)                             |
|                                                                       |
|  .   Data Transfer Services                          7. Application   |
|                                                      6. Presentation  |
|                                                      5. Session       |
|                                                      4. Transport     |
|                                                                       |
|  .   Media Layer (APEX: Carrier Networks etc)        3. Network       |
|                                                      2. Data Link     |
|                                                      1. Physical      |
|_____|
```

Figure 3. Comparison of ABC Model with ISO Model

4.Results of initial study

The requirements identified during Phase 1 of the Study are summarised below grouped into the relevant ABC layer. However a couple of general points raised during the collection of information for the first Phase of the Study are worthy of note.

Considerable difficulty was found in making the predictions of the quantity of data to be transferred in the future. It came as a surprise how small a proportion of the

Company Data is distributed electronically even in the relatively highly automated Aerospace sector of industry. This is considerably less than the amount of data created and stored on computers. It is anticipated that the quantity of data communications will increase by a factor of ten in the next five years.

The reasons for the lack of communication of data electronically even when it is available on a computer database are fundamental to the User Requirements. In summary these are:-

. lack of hardware compatibility
. difficulties in using different application packages (especially user interfaces)
. speed of access of data
. data formats incompatible between application package
. poor reliability (availability) of networks.

4.1.Information Layer

The Information and Data Layers are above the ISO Application Layer (Layer 7). Essentially the Information Layer is concerned with the logical structure of information to be exchanged and the purpose of the exchange. This is extended here beyond the APEX model to include Management and Control of the data.

Requirements of the Information Layer are for:-

. A logically integrated database of information.

. Data to be physically distributed in multiple databases (mixed database types - networked, relational, object oriented) and flat files.

. Data to be held on a number of different vendor hardware platforms.

. Ideally, data is only stored once but for efficiency or security it may be necessary to keep multiple copies of data. In these cases there should be concurrent update of all databases with database locks to prevent data getting out of synchronisation.

. Data lifetimes may be very long - up to 50 years on some aircraft projects. During this time hardware and application software will change but the data must still be accessible and usable. This may point to the desirability for all data to be held in a neutral "standard" format.

. Data changes frequently have implications on other data or users. These implications need to be identified and "alerts" or "triggers" issued. In a large organisation this could involve a considerable quantity of electronic mail being generated. The use of Knowledge Based Systems to eliminate the unnecessary messages for each user is seen as a future requirement.

. There should be an Information Management System to control:
  . user privileges for access, update etc.
  . location, storage and retrieval of data
  . configuration control (relations between data)
  . change control

and also to provide Management Information reports on activities on the data.

. Interfaces between application systems and the data should be implemented through the Information Management System with the Information Management System transparent to the general user. In order for all Application Systems to interface directly to Information Management Systems from different vendors there is a requirement for a Standard Interface Specification between Application Systems and Information Management Systems. This does not appear to have been recognised by the International Standards organisations but an Interface based on Standardised electronic messages has proved useful within British Aerospace for integrating application systems to a Management System.

4.2. Data Layer

The Data Layer handles the representation and the encoding/decoding of information being transferred. It must ensure that on receipt the coded information is decoded, verified and understandable. The information needs to be exchanged between different systems on different hardware platforms. Different encoding methods are needed for different types of information, e.g.

Text and Document
Spread sheet
Structural Alphanumeric
Image
CAD/CAM Geometry
Query Languages
Audio-Visual

There are a number of "standards" available for these types of data but the interpretation/implementation of these standards has been inconsistent between vendors and there is much room for improvement. The U.S. Department of Defence CALS (Computer-aided Acquisition and Logistic Support) (ref 4) initiative has provided a significant incentive in this area and the standards being developed in the Defence Industry will become the norm across other industrial sectors. An important part of the CALS programme, from a user's perspective, is the rigorous conformance testing and certification of products (ref 5).

The User Requirements report will review the status of the main data exchange standards and make recommendations on which standards should be adopted for implementation.

Specific requirements for the Data Layer are that:-

. data should conform to the preferred standards

. conversion of data from application native format to standard format should be efficient in terms of time and volume (one of the many problems with IGES files is the increase in data size)

. data conversion should be fully reversible. That is the data written in standard format from an application should be capable of being read back into the application with no loss of content or meaning

. data should be converted on demand into the format required by the receiving system/user. This assumes total reliability and high efficiency of data conversion. The user should be unaware of any data conversion processing.

4.3. Data Transfer Services Layer

The Data Transfer Services Layer handles the transfer of data from one user to another. It ensures the security of data from loss, theft, destruction or wrongful delivery.

Methods for data transfer are classified in a general way as:-

Store Transfer
File Transfer
Electronic Mail/Messaging
Terminal Access
Remote Job Entry
Co-operation processing
Gateways

User requirements for this layer include:-

. the message should be delivered in a timely fashion with confirmation of deliver if required

. a user should be able to retrieve any information for which he has access authority from anywhere in the system

. alarms, urgent and time critical messages must be delivered in a predetermined time span, independent of the communications system loading

. information transfers will comprise packages of very different sizes from small electronic mail messages to CAD models up to 10Mbytes in size.

The time taken to transfer an item of information will depend upon network configuration and loading and size of the information item. However the requirement of a user for delivery will vary depending on the use of the data. A large 3D model may be required immediately for reference or modification but could be transmitted overnight to a partner Company or to archive storage. Thus a means of prioritising the data is required, but this is more complex than considering the size of the data item. The priorities should be related to maximum time in transfer.

The requirement for time taken to transfer data is difficult to quantify. There is clearly a time when data arrives too slowly but very few users complain if data arrives too quickly. Often consistency of access time is as important as speed.

It is assumed that the communications system will be segmented into sub-networks in order to satisfy these requirements. In this case the router between the sub-networks should not be apparent to the user, nor should bridges, router etc. cause any delay in accessing the data.

4.4. Media Layer

The information is required to be processed on variety of hardware platforms - central computers, workstations, manufacturing machines - supplied by different vendors. Communication between these devices necessitates a communications infrastructure which enables data exchange between Sender and Receiver. This carrier may be manual (by post or courier), by telecommunications lines (Public or Private Networks or Local Area Networks (LAN)), or by radio wave (including microwave and suitable communications). The communication will most probably be to a further network via Bridges, Router, Repeaters etc.

The User Requirement Study has concentrated on the Telecommunication lines methods of transferring data. Access to the computing resources and networks should be available continuously. Reliability and speed of networks at present are inadequate to allow confident access to data on demand. Hence redundant network paths are required or more usually data is retrieved and stored for short periods (say up to 48 hours) local to the application requiring it. This is particularly the case in manufacturing areas where data requirements can be largely anticipated and where unavailability of data can cause expensive delays in terms of production machines not being used to capacity.

Network installation represents a high cost if carried out on an as required basis. Installation of a single network for a whole building or area without necessarily knowing in detail what the requirements will be in terms of attached processors, terminals, workstations or machines is far more cost effective. This assumes confidence that the equipment to be installed will interface readily to the network and hence standards for physical inter-connections and for communication protocols are universally accepted. The networks must also be over-designed and have more than sufficient capacity to accommodate future growth in data transfer. (This is estimated at a factor of ten over the next five years). The network in this terminology includes all associated hardware such as bridges, router, repeaters, gateways etc.

Message delivery should not be affected if any node is lost (unless it is the destination node for the message).

Loss of a node should be automatically detected and reported.

Removal or addition of a node should be able to be carried out with the network live without affecting data transmission to other nodes.

5. Case studies

The set of theoretical requirements has been checked against reality in two case studies within British Aerospace. These are aimed primarily at the Information and Data layers of the model rather than at the more technical levels covered by the ISO 7 Layer Model. This is probably a true reflection of the relative importance of the various layers from a User perspective. A User is generally more concerned with accessing the required information than with the type of wire or fibre used to transmit the data, except where such issues as cost, reliability, ease of installation/maintenance and the speed of data access and throughput are concerned.

The first case study considers the requirements for a Computer Aided Engineering Information Management System to enable the implementation of a Computer Integrated Business architecture ensuring the integrity and consistency of data across the CAE Functions of an Organisation. Specific objectives are:-

- manage the status of data and control of change
- release management
- associativity between data items
- audit trail
- storage/retrieval of data
- single input of data
- access/security controls on data
- control of User privileges
- open system interfaces to provideaccess to any data item from any terminal (consistent with hardware limitations)
- event triggering with Electronic Mail
- flexible query feature

Constraints on the system are the need to operate across different hardware platforms (especially DEC, IBM, UNIX and PCs). It should not be necessary for a user to know on which platform the data is situated. This will require the Information Management System to contain a register of data across all hardware platforms.

On the other hand the requirement for fast local response, and in some instances local ownership of data, will lead to the data being distributed and /or held on more than one database.

The priority is for Product Definition Data (which includes Design data, Analysis data, Manufacturing data, Test data and Quality Assurance data) to be controlled initially. This will eventually be extended to cover Life Cycle Support Data - thus matching the complete CALS Product Data Model.

Data types to be included are:-

     2D drawings
     Plot files
     3D models
     Text (Reports, Forms, Specifications)
     Calculations
     Scanned data
     N/C data
     Bills of Materials

Data types to be considered for the future are:-

     Video
     Audio
     Software
     Knowledge Base

For some data, the existing data management systems are so extensive and well established (for example Bills of Materials) that it would not be practical to consider replacing them in the short term. The CAE Information Management System must therefore interface to some existing management systems and provide a link between them.

The second Case Study is concerned with the distribution of data and ensuring that all the data required for a particular operation is available at the correct issue and status and in the desired form prior to starting the operation. It is essentially concerned with providing data associated with each Planning Operation to Production Personnel. In practice it has a wider application as it is only a means of providing a structured view of data in the CAE Information Management System. This involves the provision of:-

     a simple means of 'navigating' through the large and complex logical database

     locating and collecting the relevant data from physically distributed databases

     checking status and integrity of the data

converting the data into format desired by recipient

creating a "data kit"

distributing the data

The problems to be overcome are:-

different management and control tools operating at different levels

different data sources

broad range of recipient (people or systems) requirements

different hardware platforms

long lifespan in changing computing environment

6. Migration strategy

An important issue still to be considered in any detail is the strategy for migration from our current computing environment to the desired Open Systems environment. Any strategy must take an evolutionary approach where existing systems are in place. The choice of technology for a given environment must be based on an application specific cost/benefit analysis.

Such an analysis is likely to have the following as parameters:-

- . cost and complexity of installation
- . cost and integration level of connection hardware
- . traffic integration
- . number of connections and distances to be reached
- . real time traffic requirements and expected throughput
- . environmental factors
- . life of installation
- . stability of configuration versus evolution
- . functionality of network administration

7. Conclusions

The User Requirement Study is approximately 70% complete and will be issued at the end of 1990. It has confirmed a strong user desire for an Open Systems Computing Environment but has recognised that earlier investment in existing hardware and software may constrain the rate of progress in migrating to such an environment.

The U.S. Department of Defence CALS initiative is welcomed as a major driver towards Open Systems in the defense sector and it is anticipated that this will influence most industry sectors in the next five years.

The quantity of data to be distributed electronically will also increase by an order of magnitude over the next five years. This must be matched by the development of networking systems to operate more quickly and with greater availability than at present.

The importance of an overall Engineering Data Management system cognisant of all the data within an Organisation is becoming more widely recognised. In order for this to be fully available in an Open Computing environment standards for interfacing between Application Systems and Engineering Data Management Systems need to be developed.

8. Acknowledgements

This paper presents the results of work carried out at British Aerospace (Military Aircraft) Ltd, BAECAM, British Aerospace Computing and Telecommunications Unit, Aeritalia and the University of Porto.

The User Requirement Study is funded jointly by the ESPRIT CNMA Steering Committee and the Companies involved in the Study.

9. References

(1) ESPRIT AMICE PROJECT 5288 and 2422

(2) APEX Communication Architecture, APEX/DCA/00807 version 2, March 1990

(3) APEX Network Architecture, to be published

(4) Department of Defence, Computer-aided Acquisition and Logistic Support (CALS) Program Implementation Guide, MIL-HDBK-59A

(5) Lawrence Livermore National Laboratory, CALS Test Network - Strategic Plan

OPEN SYSTEMS PROTOCOLS AND SPECIFICATIONS

COMMUNICATIONS FOR CIM

Open Systems Protocols and Specifications

– CNMA Technical Overview –

Artur Lederhofer
SIEMENS AG, Erlangen, FRG

Karlheinz Schwarz
SIEMENS AG, Karlsruhe, FRG

Open Communication is essential for Computer Integrated Manufacturing (CIM). The ESPRIT project Communications Network for Manufacturing Applications (CNMA) aims to specify, implement, validate, and promote communication standards for manufacturing which are emerging within the framework of the International Standardization Organisation (ISO) reference model for Open Systems Interconnection (OSI).

Standardization has brought open, manufacturer independent communication in office and factories within reach. Compared with the situation only two or three years ago, considerable progress has been made, thanks not least to the success of the Manufacturing Automation Protocol (MAP) in the US and CNMA in Europe.

This presentation focuses on the technical aspects of (CNMA) for the OSI protocols used in the manufacturing environment. After an overview of the principle approach of CNMA, the applied protocol architecture and the features of the chosen networks and application protocols are described, together with the results of the specification work of CNMA. The status of the applied standards is summarized.

1. INTRODUCTION

Open Communication is essential for Computer Integrated Manufacturing (CIM). CNMA aims to specify, implement, validate, and promote communication standards for manufacturing which are emerging within the framework of the International Standards Organisation (ISO) reference model for Open Systems Interconnection (OSI Basic Reference Model, ISO 7498).

Both, administrative and control functions have been specified, realized and applied to real manufacturing applications within CNMA. Control functionality is supported by File Transfer Access and Management (FTAM) and Manufacturing Message Specification (MMS). The administration aspects are covered by Network Management (NMT) and Directory Service (DS). The features of MMS, FTAM and DS are presented, in addition to the functionality of the lower layer protocols of the seven layer stack and Local Area Network types used.

Standardization has brought open, manufacturer independent communication in office and factories within reach. Compared with the situation only two or three years ago, considerable progress has been made, thanks not least to the success of the Manufacturing Automation Protocol (MAP) in the US and CNMA in Europe.

By applying CNMA implementations to real production environments, CNMA partners have gained considerable experience which is used now to publish the CNMA results, to provide OSI products such as MAP 3.0 by CNMA vendors and to apply such products by CNMA users. The following describes the general approach of CNMA and gives some background information about ISO Base Standards, Protocol Profiles and the CNMA Implementation Guide.

2. FROM STANDARDS TO VALIDATED IMPLEMENTATIONS

The primary objectives of the CNMA Project are:

- Specification,
- Implementation,
- Validation and
- Demonstration

of Open Systems communication protocols for factory automation. Figure 1 depicts in which steps these objectives are accomplished.

a) Specification

The project strategy is to identify those communication services required to satisfy CIM user needs. The suitable services and protocols are selected from current International Standardization Organisation specifications. CNMA protocols and features are specified in an Implementation Guide (IG). This guide is the common basis for the implementations of the CNMA protocols.

While specifying the CNMA Implementation Guide, specific CNMA user requirements had to be taken into account for functionality needed of the various CNMA pilot applications.

b) Implementation

Proof of the specification work comes only from implementations. Implementation results are fed back into the CNMA specification work and the standardization process at national and international level.

c) Validation

All CNMA implementations have been conformance tested and tested for interoperability at a common test platform, the "Institut für Steuerungs- und Werkzeugmaschinen" (ISW) at the University of Stuttgart. After these tests application software for the ISW pilot was commissioned.

Further proof of the stability of the communication software will be gained at the other pilots of CNMA within real production environments.

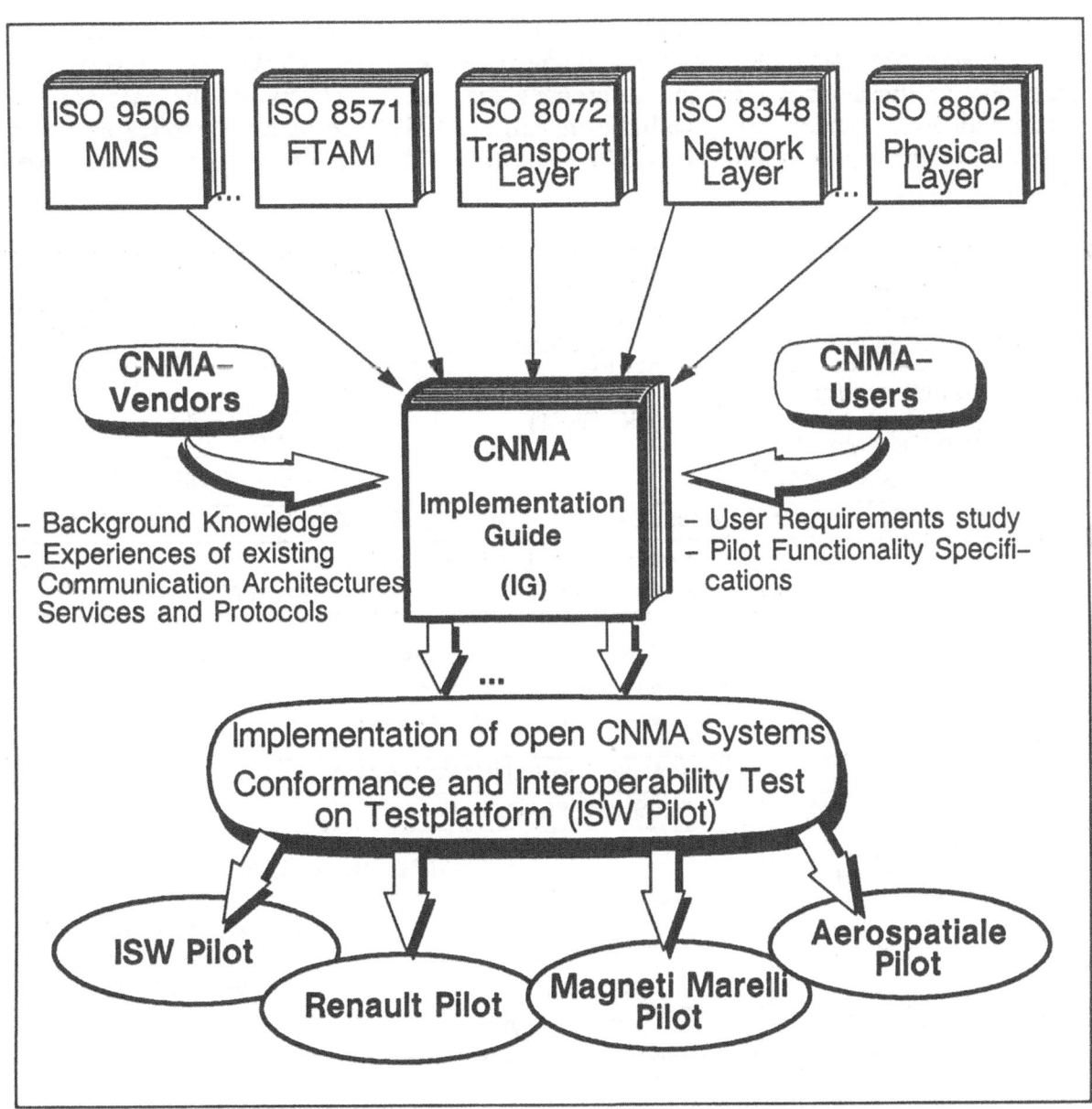

Figure 1: CNMA Approach

d) Demonstration

In order to promote the Open System Interconnection principles and to give proof of successful implementations CNMA partners believe that significant public demonstrations of the work are vital. By such demonstrations it is possible to promote the development of internationally agreed protocols for manufacturing applications and to disseminate the project results.

The ISW pilot is open to the public during the AMB fair where life demonstrations are given. CNMA pilots of Renault (France), Aerosptiale (France), and Magnetic Marelli (Italy) will be also open for the public after commissioning in autumn 1990.

3. CNMA SPECIFICATION – IMPLEMENTATION GUIDE

The ISO Reference Model for Open Systems Interconnection (OSI) [1] has been defined to form a framework for the development of communication protocol standards. Layer 1 to 4 cover reliable data transmission with error detection and correction, and layers 5, 6 and 7 govern the application oriented dialog between users. CNMA has been designed according to those principles and references the latest versions of international standards and drafts which were available in August 1989, the date of publication of the CNMA Implementation Guide for CNMA Phase IV. The chosen protocol architecture is shown in figure 2.

7	MMS Companion Standards 1) (ISO CD 9506/4 and /5)	File Transfer Access and Management, FTAM (ISO 8571)	Networkmanagement			Directory Service (ISO 9594)
			CM ISO CD 10164-1	NMT PM WD N3313	FM WD N3312	
	Manufacturing Message Specification, MMS (ISO 9506/1 and /2)		CMIS/CMIP (ISO DIS 9595-2/9596-2)			
			ROSE (ISO DIS 9072)			
	Association Control Service Element, ACSE, (ISO DIS 8649/8650, N2526, N2327)					
6	Presentation (ISO 8822/8823) Kernel Abstract Syntax Notation One, ASN.1, (ISO 8824/8825)					
5	Session (ISO 8326/8327) Kernel, Full Duplex, Session Version 1/2					
4	Transport (ISO 8072/8073) Class 4					
3	Connectionless Internet (ISO 8348/8473)					
	PLP (CCITT X.25)	ES/IS (ISO 9542), optional				
2	HDLC LAP B (CCITT X.25)	LLC 1 (ISO DIS 8802/2)				
1	X.21/X.21 bis	CSMA/CD 10 MBit/s (ISO 8802/3)	Token Bus (ISO DIS 8802/4) Broadband 10 MBit/s Carrierband 5 MBit/s			
	MAP, TOP, CNMA	TOP, CNMA	MAP, CNMA			

1) oriented on early working drafts

Figure 2: CNMA Phase IV Protocol Architecture

As can be seen from figure 3, an OSI protocol architecture comprises a series of refe-renced ISO Standards, so called Base Standards which sometimes contain much more functionality than needed in particular applications. What is needed is a protocol profile which defines which Base Standards, options and parameters are needed. The CNMA Implementation Guides [2] contain exact protocol profiles for each project phase. Using these profiles, vendors develop the protocol software and system interfaces for their equipment (programmable controllers, NCs, minicomputers) in order to evaluate imple-mentations, to gather experience, to uncover loopholes in the standards, and to find ways to close them.

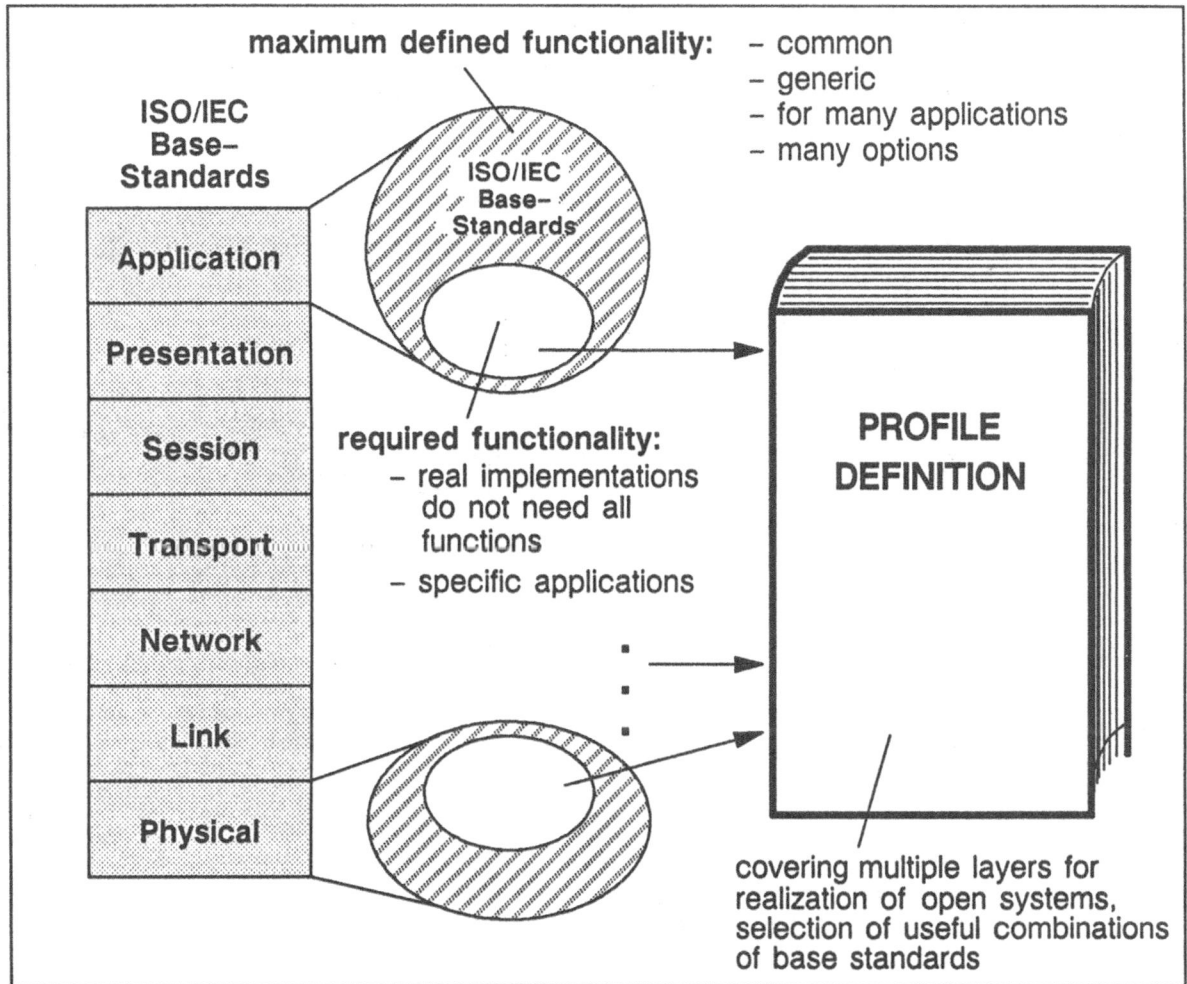

Figure 3: Profile definition

The CNMA Implementation Guide ist divided into three Volumes:

CNMA Implementation Guide Volume 1 (T-Profile)

Volume 1 is a Transport Profile (T-Profile) definition which caters for OSI layers 1-4. This definition had to take into account CNMA Vendor's background implementations, since no new implementation work was done on these layers within the scope of CNMA. The specification of this profile stays in line with european prestandards on single LAN [3] and multiple LAN [4].

CNMA Implementation Guide Volume 2 (A–Profile)

Volume 2 is an Application Profile (A–Profile) definition for multiple application protocols comprising OSI layers 5–7. The specific OSI application protocols addressed are:

- Manufacturing Message Specification (MMS)
- Directory Service (DS)
- Network Management (NMT)
- File Transfer, Access and Management (FTAM)

Figure 4 presents the structure of this volume. Since application protocols build on top of lower OSI protocols, commonly used functionality has been defined in selfstanding parts of the document. The structure of this volume was set up according to the guidelines of ISO/IEC TR 10000 [5] which defines the required structure of International Standardized Profiles (ISPs).

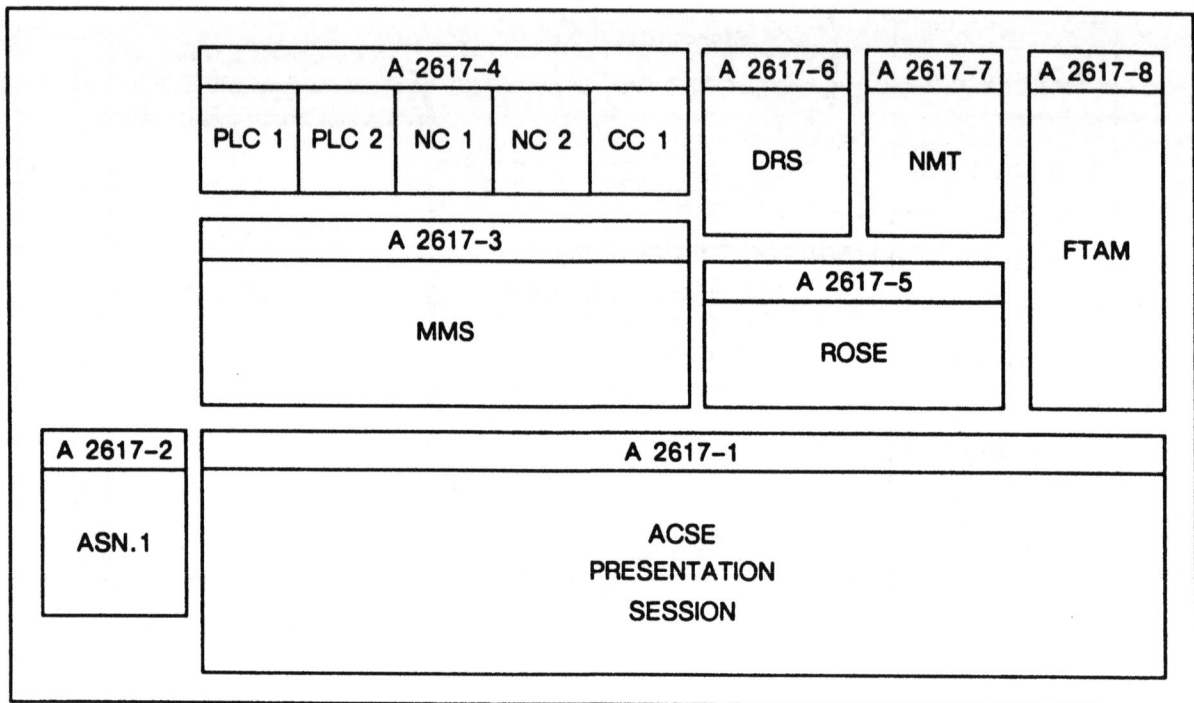

Figure 4: Structure of CNMA Implementation Guide Volume 2

CNMA Implementation Guide Volume 3 (Standardized Interfaces)

In order to allow easy porting of application software written for host computers, such as minis and personal computers, CNMA specifies standardized interfaces for MMS and FTAM.

Changes of the underlying application protocol can be hidden to the application and the so called "High Level" interface frees the application programmer from the underlying details of the application protocol. These interfaces are currently defined for "C"–language bindings, within Volume 3 of the Implementation Guide. The interface specifications are derived from valuable work done within MAP 3.0 [8]. These specifications have been adopted within CNMA to match subsets chosen for MMS and FTAM.

4. FUNCTIONALITY OF SERVICES AND PROTOCOLS

Automation technology for shop–floor and technical office applications consists of a variety of functionally staged components which are interconnected via a communications network, to form the basis for Computer Integrated Manufacturing (CIM). Such a network is made–up of various controllers (PLCs, robot controls, numerical controls, drive controls, etc.), cell control computers (minicomputers or PC), workstations or PCs for programming and document handling, mainframes for order handling, accounting and corporate planning. An example is shown in figure 5.

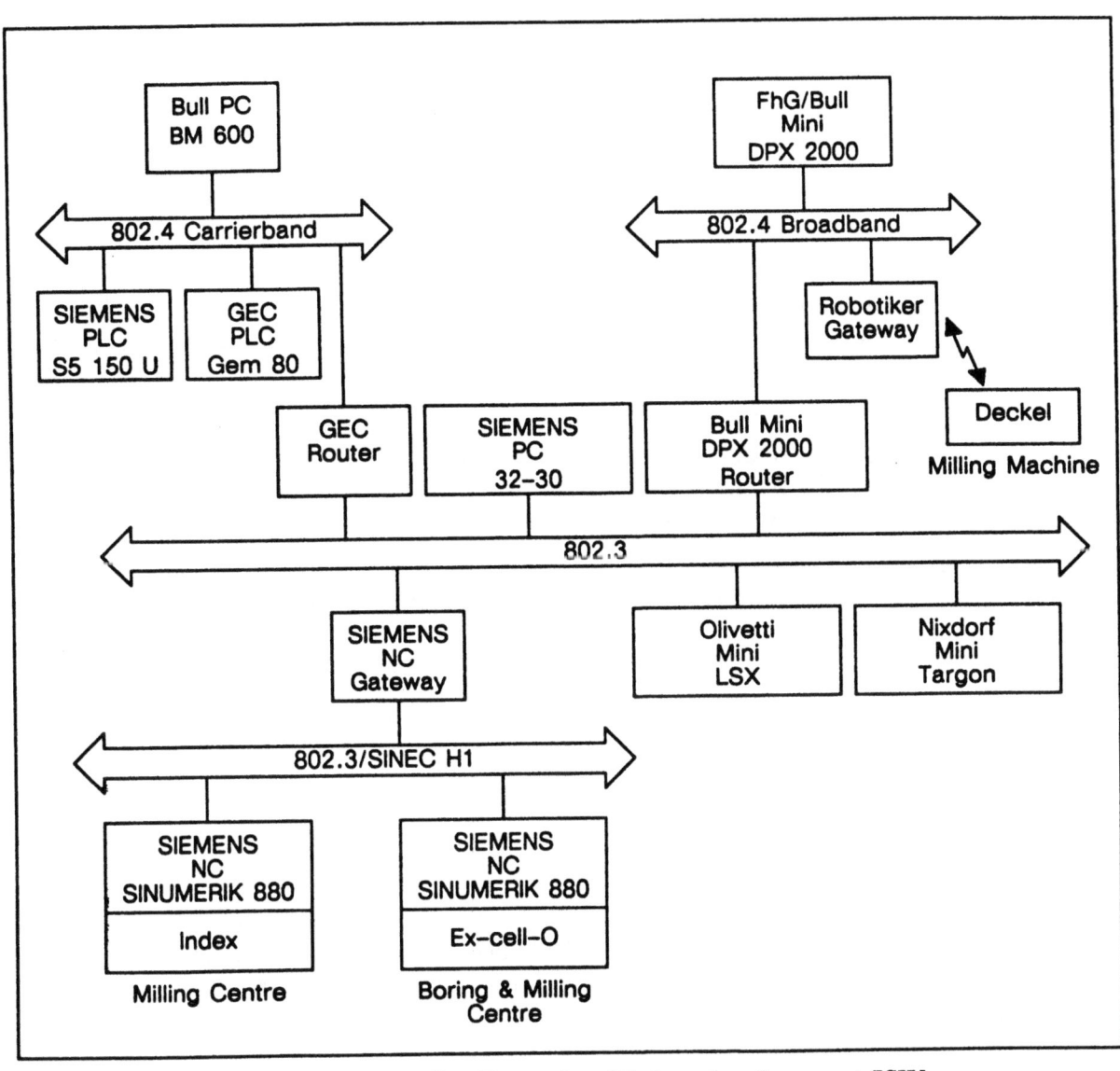

Figure 5: Pilot Configuration University Stuttgart ISW

By applying and implementing OSI standards, interconnectivity of these highly heterogeneous components in terms of hardware, system software (i.e. operating systems) and application software can be achieved in a standardized, homogeneous way. Options for implementations are provided for the application layer services and the network types. These are of concern and apparent to the user since they address questions like these: What Networks can I use? What application services do I get?

4.1 NETWORKS

CNMA currently uses the following Local Area Networks as means to link devices:

1. IEEE 802.3: Carrier Sense Multiple Access with Collision Detection (CSMA/CD), operating at 10 MBit per second with 500 m cable segments.

2. IEEE 802.4: Token Bus Broadband, operating at 10 Mbit per second, modulated onto a broadband cable system which can be shared with other technologies such as cable TV, terminal multiplexers etc.; this technology facilitates as backbone media cable system up to and beyond 10 km in length.

3. IEEE 802.4: Token Bus Carrierband, operating at 5 Mbit per second; typical carrierband systems may use cable lengths between 50 m and 700 m depending on number of stations and trunk cable type.

In addition, access to remote sides is necessary via Wide Area Networks. CNMA therefore caters also for X.25 links.

Providing a range of networks allows to select the network which most suits the user's implementation requirements. Requirement considerations will include: the topology which has to be constructed; whether services have to share the same cable; the geographic area to be spanned; the installed base; cost and preferences; environment; etc.

CNMA allows any of its selected LANs to be combined to form a single logical network. This means that a device on one LAN can communicate with a device on another LAN: LAN's are interconnected via bridges or routers.

Bridges perform the interconnection of the LANs on layer 2, above the medium access control (MAC), while forming one logical subnet in terms of administration of stations on the interconnected LANs.

Routers may have two or more ports, one on each subnet. Routing is performed by receiving data on layer 3 from one subnetwork and transmitting it to another. Routers have been used within CNMA to also provide connections to remote sides via X.25 public packet networks.

4.2 APPLICATION LAYER

At layer 7 of the OSI reference model CNMA offers implementations for different application layer standards: for MMS, FTAM, Network Management and Directory Service. The functionality of these protocols, except for Network Management are described in the following sections.

4.2.1 MANUFACTURING MESSAGE SPECIFICATION (MMS)

The Manufacturing Message Specification (MMS) is the most important application layer standard in the area of industrial automation. MMS provides a wide variety of communication services useful for any kind of controlling devices in a distributed system environment for discrete parts manufacturing and process control applications. The aim of MMS is to allow, with a minimum knowledge of internals of the remote device, the interconnection of a local with a remote system.

MMS supports communications between programmable devices such as Robot Controllers (RCs), Numerical Controllers (NCs), Programmable Logic Controllers (PLCs) and other intelligent devices e.g. Cell Controllers. The MMS Specification is referenced by MAP 3.0 Specification and CNMA Implementation Guide as the primary Application Layer Protocol for the factory floor.

4.2.1.1 INTRODUCTION TO MMS MODELS AND SERVICES

The basic concepts of MMS are the so-called Virtual Manufacturing Device (VMD) and the Client-Server-Model. The overall modelling of MMS is shown in figure 6 which depicts two devices connected by a communication system to ISO/OSI standards.

One user plays the client role, requesting another device (the server) to perform some application-specific operation. The request is transferred by an Request Protocol Data Unit (PDU). The other plays the MMS server role, performing the requested operation and responding with information resulting from the operation. The Response is transfered by an Response PDU.

A VMD – defined in the Server – represents the standardized view of the structure and external visible behaviour of real manufacturing devices and makes available, for control and monitoring, the resources and functionality associated with a real manufacturing device. Within the scope of a VMD, MMS defines 15 objects and about 80 operations on these objects.

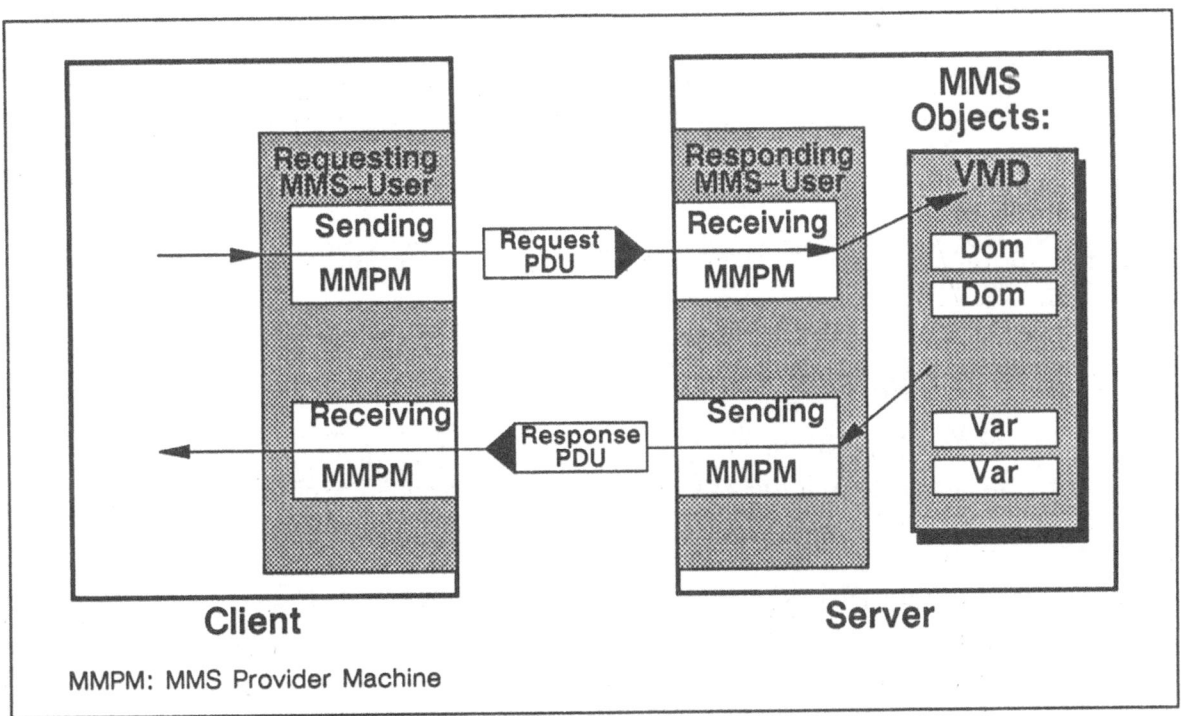

Figure 6: MMS Client-Server-Modell

4.2.1.2 MMS OBJECTS AND SERVICES

The VMD contains MMS objects, which are made available for manipulation by MMS services. Such objects include for example variables, domains, program invocations, semaphores and event actions. A short description of these objects is given below (the names in the parenthesis describe examples of services which can be executed):

Transaction Object – State information of a confirmed service.

Domain – A Load image containing programs or data (Download, Upload, Delete, ... of Domains).

Program Invocation – An executable program consisting of one or more Domains (Start, Stop, Resume, Kill, Delete, ... of Program Invocations).

Variable, Variable List, Scattered Access – Comparable with variable access in high-level computer languages (Read, Write, Information Report, Define Variable, ...).

Type – A data type useful for accessing MMS variables.

Semaphore – A flag indicating if a resource is beeing used or free; for exclusive access or synchronisation of MMS users (Take Control, Relinguish Control, ...).

Event Condition – A logical condition as seen by the network (Define, Delete, Alter Event Condition Monitoring, ...).

Event Action – An action (MMS Service) to be executed by the server when an event occurs (Define, Delete, ...).

Event Enrollment – Enrolls a remote client for receiving an Event Notification (Define, Get Alarm Summary, Event Notification, Acknowledge Event Notification, ...).

Journal – A log of information for tracking activity on the factory floor (Write Journal, Read Journal, ...).

Operator Station – A console at which a human operator sends and receives messages (Input and Output).

4.2.1.3 MMS COMPANION STANDARDS

The services in MMS represent the common or generic services that can be applied to a wide range of devices. In some cases, however, there are specialized, device-specific parameters set aside for so-called Companion Standards, giving further details as to how MMS is to be used with specific types of devices.

Each Companion Standard defines a model of the application to be modelled by MMS and describes the mapping of the modelled functionality on a subset of MMS models, services and parameters. Companion Standards are currently beeing defined for robot controllers, numerical controllers, programmable logic controllers, process industries and production management (figure 7 lists the assigned number and status of the Companion Standards).

Companion Standard definitions cover the following issues:

● Mapping of application specific models on MMS models,

● Application specific enhancements,

● Predefined objects (all attributes defined, support or non support of options),

● Standardized names for objects,

● Subsetting of MMS services, options and parameters and

● Augmentation of application specific services, options and parameters.

When CNMA Phase IV started, Companion Standards were in a very early stage. Therefore, CNMA currently uses MMS part 1 and part 2 and takes into account the basic spirit of the respective Companion Standards for PLC and NC devices. As long as stable Companion Standard are missing, MMS can be used <u>independent</u> of Companion Standards. This has been shown by the results of CNMA and MAP 3.0 where the MMS core has been used without definition of new semantics and syntax of the protocol (i.e. MMS part 1 and part 2).

```
MMS  ISO 9506:
 - Services                          (Part 1)    International Standard (IS)
 - Protocol                          (Part 2)    International Standard (IS)

Companion Standard:
 - for Robots                        (Part 3)    Draft International Standard (DIS)
 - for Numerical Control             (Part 4)    Committee Darft (CD)
 - for Programmable Controllers      (Part 5)    Committee Draft (CD)
 - for Process Industries            (Part 6)    Committee Draft (CD)
 - for Production Management         (Part 7)    Work just starting
 ...
```

Figure 7: Stability of MMS and Companion Standards

By applying MMS to real manufacturing devices and feeding back the experience gained into the standardization of Companion Standards, CNMA believes to make the standardization of Companion Standards most productive. CNMA anticipates to reference and apply the evolved Companion Standards in the future.

4.2.1.4 CNMA SUBSET OF MMS

CNMA has selected a useful subset of MMS Models, Objects, Services and Parameters to solve the required functionality for the various CNMA pilots. The selection of this subset was gained by taking also into account MAP 3.0 (MAP 3.0 defines 7 conformance classes: MAP 1 to MAP 7) and early Companion Standards for Numerical Control and Programmable Controllers.

Five Conformance classes are defined:

- NC 1   For simple Numeric Controllers, this class is below MAP 1. Used Services: Environment and General Management Services, all Domain Services, Input and Output.

- NC 2   For more complex Numeric Controllers, this class is a superset of MAP 2 and MAP 3. Used Services: Environment and General Management Services, all Domain Services, Read, Write, Information Report, Get VariableAccessAttributes, all Program Invocation Services, Input, Output, and Event Notification.

- PLC 1   For simple Programmable Controller, this class is a superset of MAP 2. Used Services: Environment and General Management Services, Read, Write, Information Report and Get VariableAccessAttributes.

- PLC 2   For more complex Programmable Controller, this class is identical NC 2 with the exeption of Input and Output which are not present in PLC 2, and is a superset of MAP 3.

- CC   For Cell Controller applications, this class is identical PLC 1 and provides in addition FileOpen, FileRead, FileClose, and ObtainFile.

Figure 8 depicts the service subsetting of MMS as defined in the CNMA Implementation Guide.The selection of the subset shown fulfills the current needs of CNMA pilots. It is anticipated for the future that CNMA will use the subsets that will be defined in Companion Standards and in the area of International Standardized Profiles.

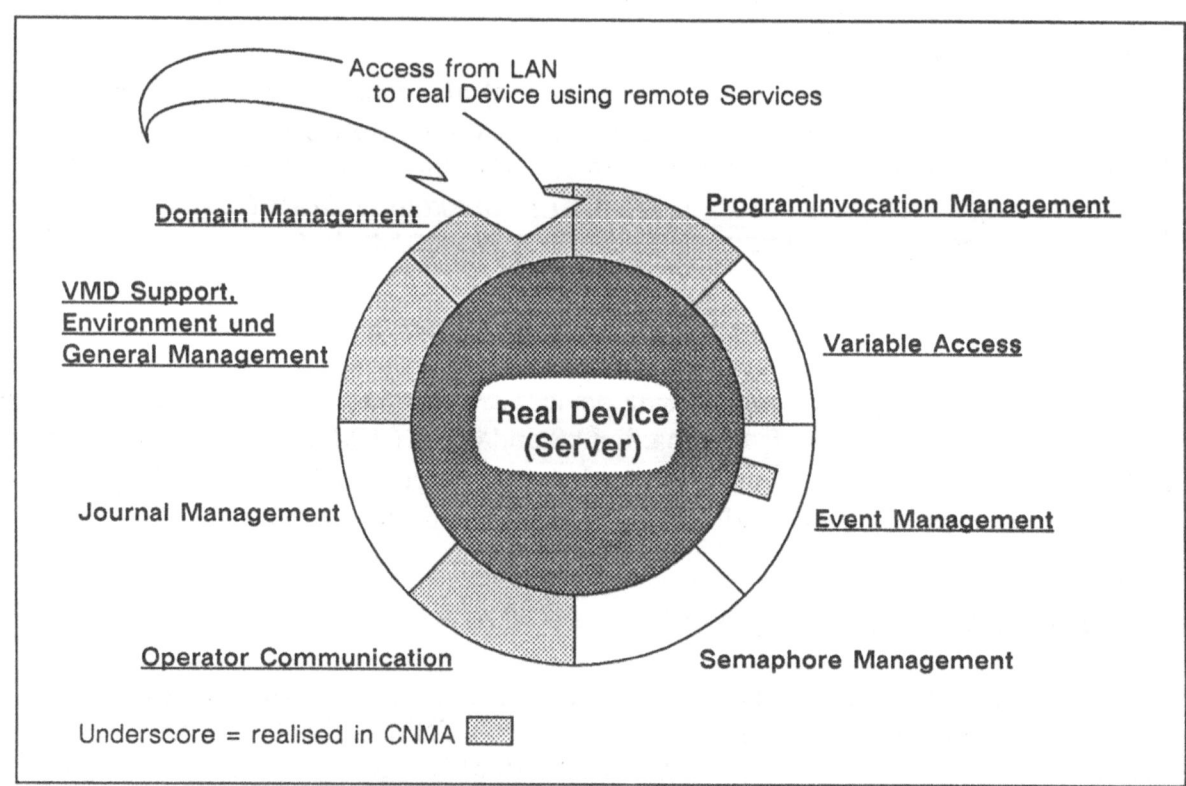

Figure 8: CNMA Service Subsetting

4.2.2 FTAM

FTAM, the application service for file transfer, access and management supports the transmission of various types of files within a heterogeneous environment. FTAM is mainly targeted for systems with peripheral filestores. Therefore, within CNMA, FTAM is to be used for computer to computer communications at factory backbone level (i.e. among cell controllers, workstations, PCs) rather than for communications between cell controllers and controlling devices.

The basic concept of FTAM is the definition of a Virtual Filestore on which the different operations can be performed. The virtual filestore defines a universal file model which can be mapped to a variety of real file types and file access methods. The virtual filestore definition includes the description of a general hierarchical access model to specify the file's contents and the access paths to the data units within the file. For simple files, FTAM permits the definition of constraint sets to simplify the general model and to reduce the number of possible operations (for this model see figure 9).

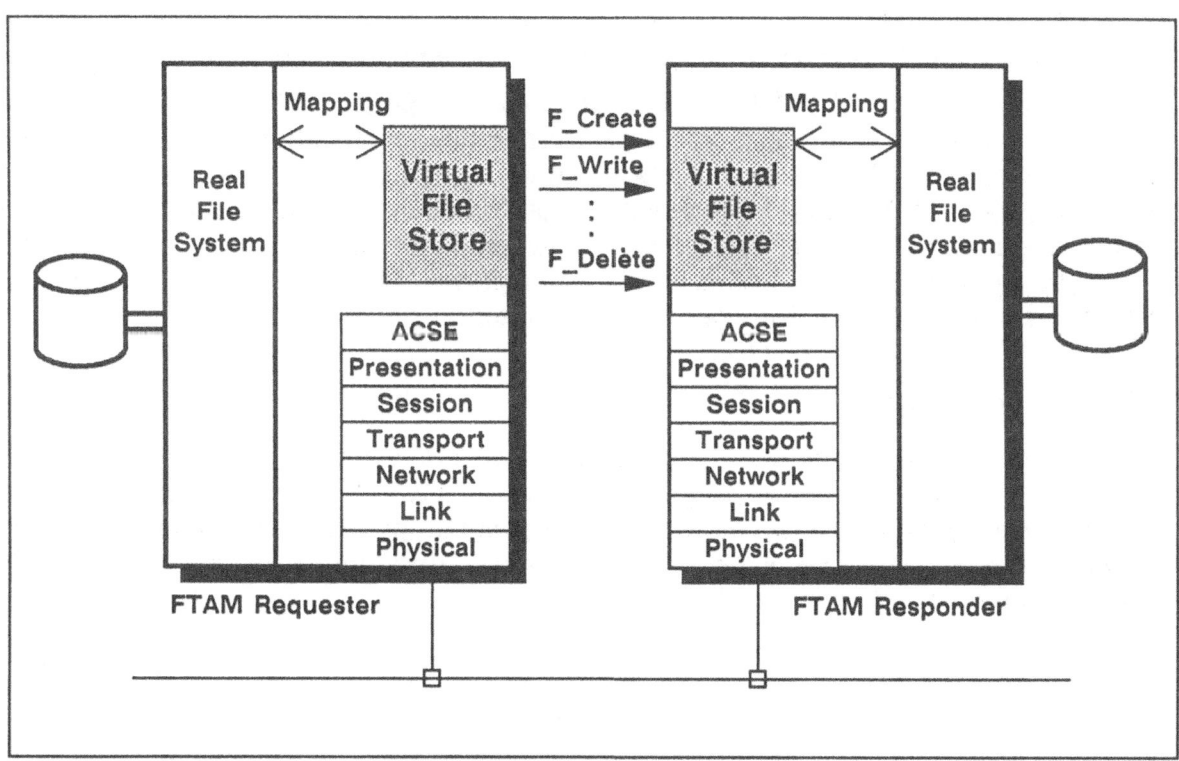

Figure 9: FTAM Model

The internal structure of a file must be maintained during transmission. For the purpose of transfer, a file's contents may be described by means of an abstract syntax. A file may be accessed (or transferred) in different access contexts depending on the level of detail required for structuring information (e.g. record layout). To maintian the file's properties during transmission, both end systems need to have a common understanding of the file specific parameters. Therefore, each file is associated with a document type. The document type serves as a repository for file parameters important for storage and transmission.

CNMA FTAM implementations are capable of acting as both, initiator and responder of FTAM services to

- establish, terminate or abort an FTAM regime (i.e., an FTAM association)
- select and deselect a file
- open and close a file
- read and write from/to a file
- to cancel a bulk data transfer
- create and delete a file
- read the attributes of a file.

During the establishment of an FTAM regime, login parameters such as "Login–Name", "Password" and "Account" can be exchanged for proper setup of the FTAM responder's environment.

The FTAM service subset selected permits the complete transmission of text files (document type FTAM 1) and binary files (FTAM 3) in both directions, i.e. copy a file from the local to a remote filestore or copy a file from a remote to the local filestore.

Since well established and widely agreed functional standards/profiles (ENV 41 204 [6] and NIST [7] which is referenced in the MAP/TOP 3.0 specification) are already available for the functionality needed, CNMA rather references those profiles instead of defining a new profile. FTAM functionality for CNMA is limited to file transfer, i.e. transfer of complete files, and limited file management as defined in [6] and NIST/OSI [7]. No checkpointing, restart and recovery is supported. Only files with simple internal structures, i.e. unstructured files can be transmitted.

4.2.3 Directory Service

The Directory is a facility which supports the storage and interrogation of information about named objects (things or people), in order to provide services such as network access to "White Pages" (to extract information from an explicitly identified object) and to "Yellow Pages" (to obtain a list of subordinate information of an identified object). Figure 10 depicts the Directory Model.

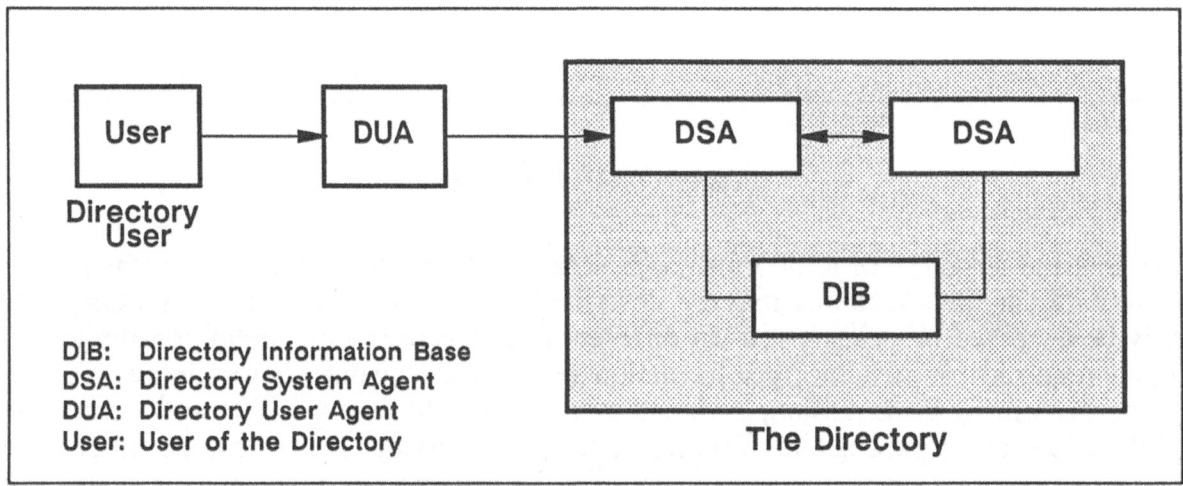

Figure 10: Directory Model

The information held by the Directory, collectively known as the Directory Information Base (DIB), is typically used to facilitate communication between, with or about objects such as application entities, people, terminals and distribution lists.

The DIB is composed of directory entries. A directory entry consists of a set of attributes each having one ore more values. Some of these attributes are mandatory, while others are optional. There are a number of attribute types which the Directory knows about and uses for its own purposes. They include the attribute "object class" which appears in every entry and indicates the object class to which the object belongs.

The entries in the DIB are arranged in the form of a tree, the Directory Information Tree (DIT), where the vertices represent the entries. Each entry has a distinguished name which uniquely and unambiguously identifies it. The distinguished name of an entry is made up of the distinguished name of its immediate superior entry, together with specially nominated attribute values from the entry (these attributes uniquely identify that entry and form the relative distinguished name). Some of the entries at the leaves of the tree are alias entries, they point to object entries and provide the basis for alternative naming for the corresponding objects.

From a functional point of view the Directory is a set of one or more Directory System Agents (DSAs) each providing zero, one ore more access points. A centralized Directory is a special case of this model, namely where there is only one DSA. Where the Directory is composed of more than one DSA, it is said to be distributed and the DSAs, when necessary, communicate with each other in order to satisfy the request of the Directory User.

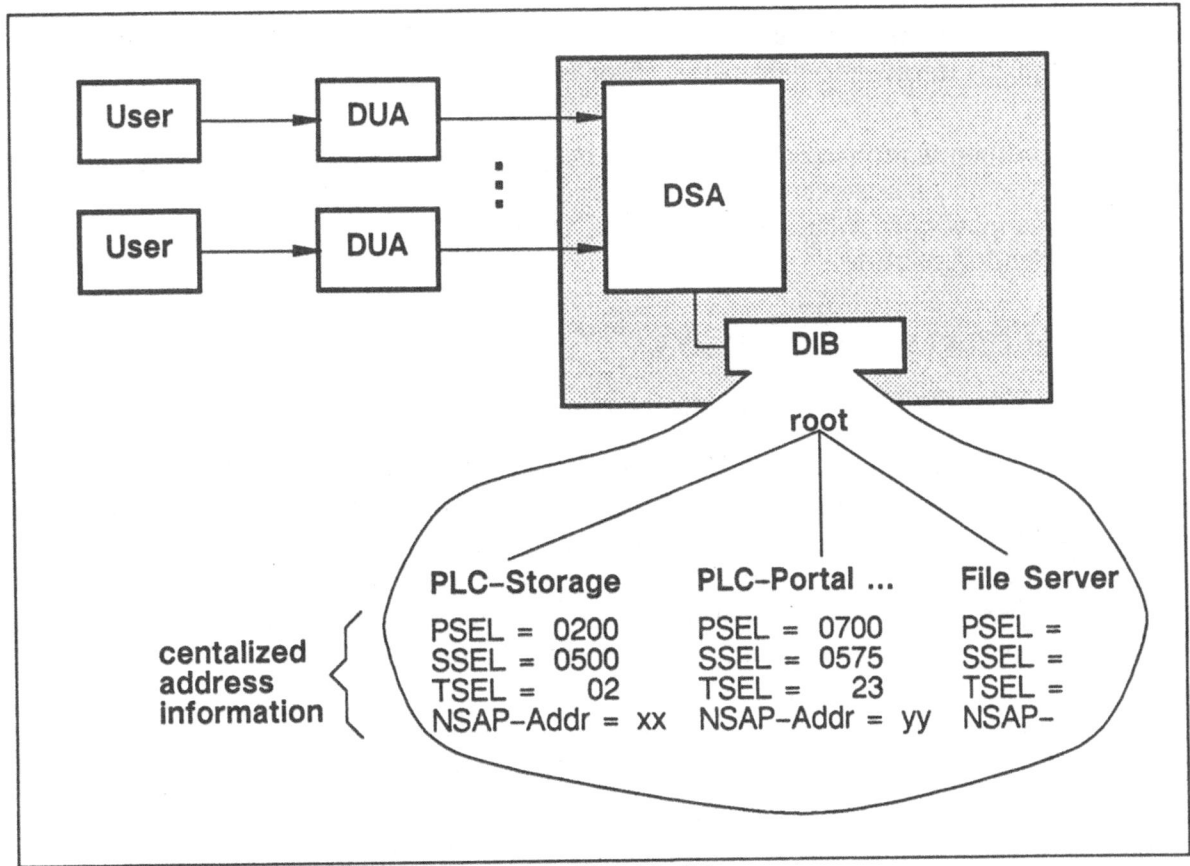

Figure 11: CNMA Directory

Two protocols are provided, namely the Directory Access Protocol (DAP), which is used for the communication between a DUA and a DSA, and the Directory System Protocol (DSP), which is used for the communication between DSAs.

Directory User may be a person or an application–process. The Directory User accesses the Directory using services available from a Directory User Agent (DUA).

Based on needs within industrial environments and taking into account limitations of controllers like PLC, NC, robot control, CNMA assumes a centralized directory. This requires only one DSA which holds the DIB (see figure 11).

The application of the Directory in CNMA solves the problem of name to address mapping and provides a central repository of the address information. Address details of almost each ISO layer is needed when establishing a connection between application entities. If no central name to address mapping is available, each individual endsystem will need to maintain all address details locally.

The DUAs reside in the end users' systems and handle the protocol for communication between a Directory user and a DSA in order to access the directory.

Three types of operations are provided, namely operations for the establishment and the release of an association (Bind and Unbind operations), operations for accessing the information contained in the DIB (Read, Compare and Abandon operations) and operations for the manipulation of the stored information (AddEntry, RemoveEntry, ModifyEntry and ModifyRDN operations).

For CNMA only the operations required for the support of the "White Pages" facility are provided. Two kinds of users are identified, on the basis of the authority required to perform certain operations on the directory, namely the administrator, which is responsible for management of data (i.e. can modify the DIB) and the interrogator which simply requires information from the Directory (i.e. can only "read" the information in the DIB).

A simple authentication mechanism, enhancing protection against unauthorized access and the basis for secure services is provided. Each user has to provide his name and optionally a password when accessing the Directory with the Bind operation, for the purpose of administration.

4. STATUS OF STANDARDIZATION

Progress and acceptance of open communication architectures have suffered from the lack of mature and stable protocol standards in the past. Tremendous progress has been made in recent years in the standardization area, since the necessity for manufacturer independent communication has been widely recognized. This progress was helped significantly by the MAP/TOP projects in the U.S. and CNMA in Europe. In addition to the progression of the base standards – primarily developed by ISO, IEEE, IEC and CCITT – vendor and user groups have been established to agree on achievable subsets of the base standards and to develop Functional Standards or Profiles.

An obvious prerequisite for the provision of OSI products is the existence of stable standards. In ISO/IEC the status of individual standards is defined by the stages Working Draft (WD) / Committee Draft (CD) / Draft International Standard (DIS) / International Standard (IS). Our experience so far shows that up to CD the changes of standards may lead to complete reimplementations and major changes of the implementation architecture. After CD, greater changes to the communications software take place which may well affect the interface. Stability for the communications software is only really assured after IS.

Figure 12 shows history and forecasts for the further development of important standards for factory automation. First products can be expected approximately one year after publication of the standards (IS). If we require a complete profile on the basis of IS including Companion Standards and NMT, <u>broad</u> application may not become reality until the mid-90s.

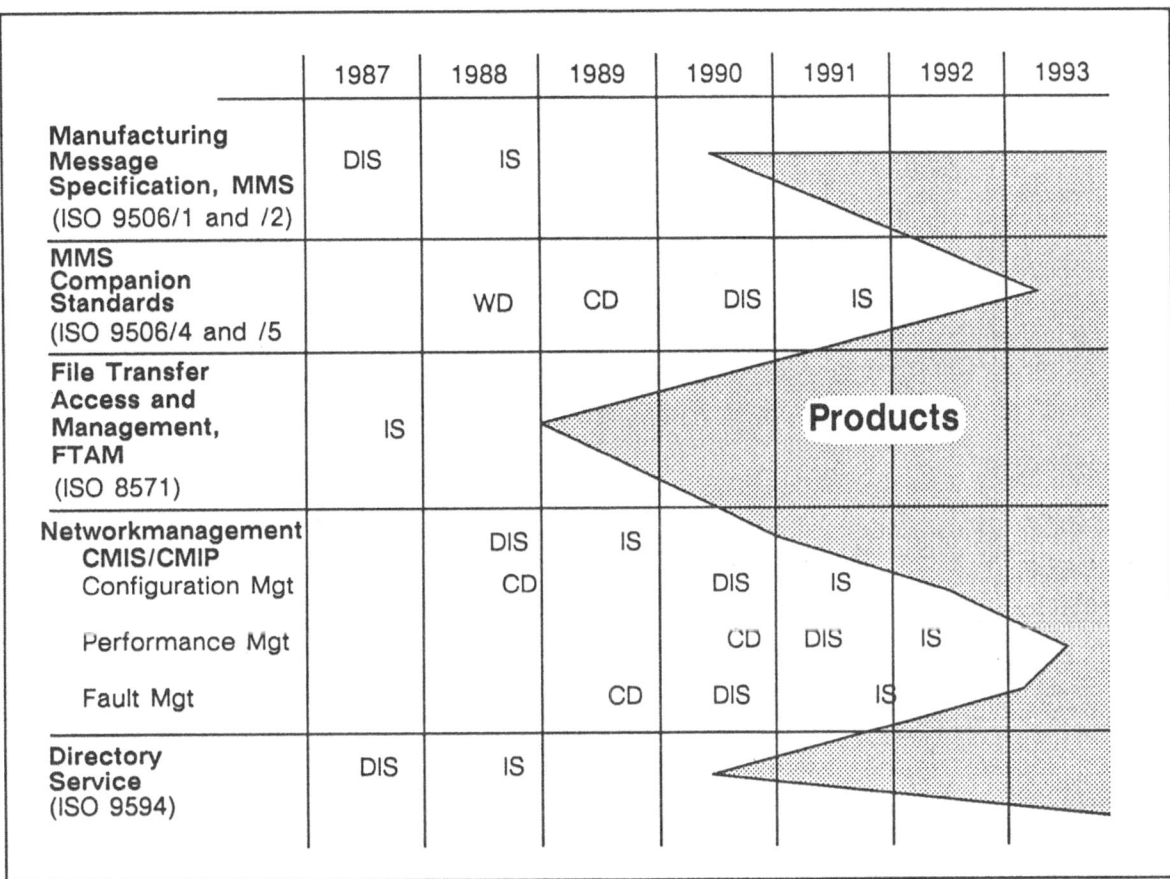

Figure 12: Status of Standardization

CNMA participating members are active in standardization of ISO/IEC base standards through national Standardization bodies such as DIN, AFNOR, BSI, etc. to give feedback of implementation experience.

In addition, liasons have been established to the regional workshops (see figure 13) responsible for production of proposals for International Standardized Profiles:

- European Workshop (EWOS) in Europe and
- Open Implementor's Workshop at the National Institue of Standards and Technology (NIST-OIW) in the U.S.

Within those workshops International Standardized Profiles will be proposed which will replace current de facto profiles such as MAP/TOP 3.0 and the CNMA IG in longterm view. The three regional workshops EWOS, NIST-OIW and the Asean/Oceanic Workshop (AOW) harmonize their ISP proposals before they are endorsed by ISO/IEC as final ISPs. As those International Standardized Profiles stabilize, it is intended to reference those within future CNMA IGs rather than defining own ones.

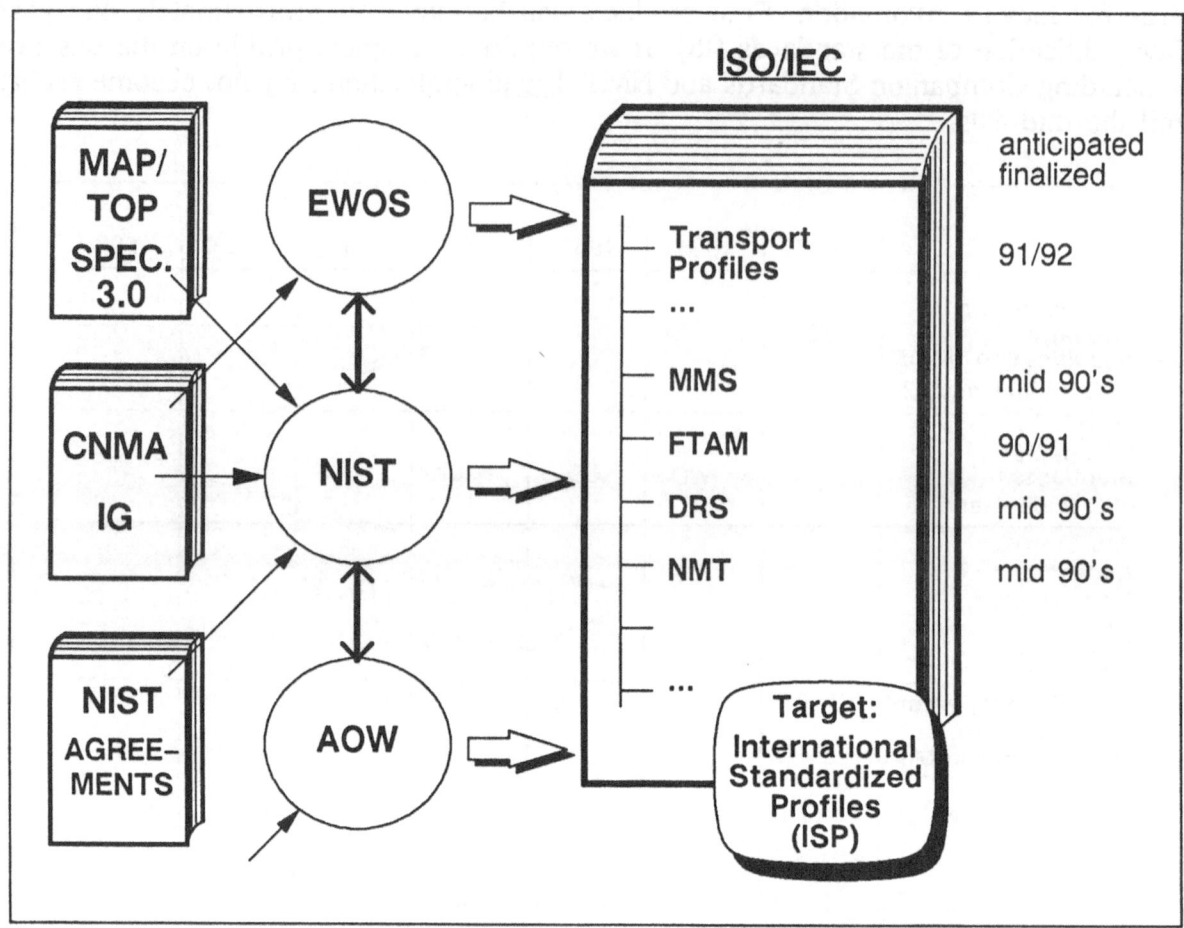

Figure 13: International Standardized Profiles (ISP)

5. CONCLUSION

o Open Communication is a basic prerequisite on the long way to CIM. Pivotal points of open communications are stable, longlived applied standards.

o Projects like CNMA including vendors and users were (and are still) necessary to:

- identify and specify user needs

- define a vendor-independent Implementation Guide which is based on the state-of-the-art in OSI communication and which provides a consistent base for implementations

- develop independant test tools in order to validate implementations against the agreed specification

- develop communication software to connect the controllers and computers of the respective vendors to the OSI networks

- use these implementations in real production facilities in order to

- prove that the user requirements have been met

- provide the user with the opportunity to get experiences with OSI communication technology in their own factories

- promote the results through open demonstrators

- feed the results into standardization in order to come to an internationally agreed and stable set of standards of OSI communication in CIM.

o The complexity of the communication technology must not be underestimated; this is due to the rich functionality of mostly all layers of the OSI Protocols.

o It has been experienced during commissioning of CNMA communications software, that IEEE 802.3 LANs were easier to set up and maintain than 802.4 LANs. IEEE 802.4 Medium Access Control Parameters need to be clearly understood when tuning the network.

o LAN Analysers to monitor, filter and display all traffic during commissioning of the networks have been found to be a necessity. They are of particular benefit if they are able to display upper ISO Layer Protocol Data in easy readable format.

o There are still areas which need further work within the next few years (especially MMS Companion Standards, Network Management and International Standardized Profiles).

o But in contrary to the situation in the second half of the 1980's (where technology and market were premature), it is now evident that Open Communication in the factory is leaving the status of prototypes and pilot implementations. More and more products appear on the marketplace and users (especially in the automation industry) are planning OSI installations now.

o One key issue for a successful application of OSI standards is that vendors offer clear strategies for the migration from their proprietary protocols to the OSI protocols.

6. REFERENCES

[1]   ISO 7498 – Information Processing Systems – Open Systems Interconnection – Basic Reference Model, 1984

[2]   CNMA Implementation Guide V.4.1, 1990

[3]   CEN/CENELEC ENV 41 101 European Prestandard – Information Systems Interconnection – Local Area Network – Provision of the OSI Connection-mode Transport Service using Connectionless–mode Network Service on a CSMA/CD single LAN environment (June 1986)

[4]   CEN/CENELEC ENV 41 102 European Prestandard – Information Systems Interconnection – Local Area Network – Provision of the OSI Connection-mode Transport Service and the OSI Connectionless–mode Network Service on a CSMA/CD single or mulitple LAN environment (June 1986)

[5]   ISO/IEC/TR 10000 Information Processing Systems – Open System Interconnection – Information Technology Framework and Taxonomy of International Standardized Profiles – Technical Report, Part 1: Framework (ISO/IEC JTC1/SGFS N 184, 1990-02-09).

[6]   CEN/CENELEC ENV 41 204 European Prestandard – Information Systems Interconnection; File Transfer, Access and Management; Simple Filetransfer (June 1988)

[7]   NIST Special Publication 500-162 – Stable Implementation Agreements for Open Systems Interconnection Protocols (Version 2, Edition 1, Dec. 88).

[8]   Manufacturing Automation Protocol (MAP) – Specification Version 3.0, August 1988.

ISW PILOT

Gerhard Krebser (ISW), Herbert Lauerer (ISW)

ISW
Universität Stuttgart
Seidenstrasse 36
7000 Stuttgart 1
Germany

Summary

The "Institut für Steuerungstechnik der Werkzeugmaschinen und Fertigungseinsichuungen" of the University of Stuttgart (ISW) has provided an experimental pilot facility, where the benefits of open systems communications are demonstrated. The pilot uses devices (minicomputers, PC's and controllers) from each participating vendor within the CNMA project and by this way represents a real multivendor CIM environment. It is divided into two independant parts. One part shows a fully automated production of discrete workpieces using two machine tool centres, which are both connected to a pallet store with a linear portal robot. The other part shows a link between CAD, CAP and NC based on CNMA's communication technology using different minicomputers for CAD and CAP and a 5-axis milling machine.

1 Introduction

One of the most important topics in the area of computer integrated manufacturing (CIM) is the communication between computers and controllers interconnected through local area networks (LAN). A burning problem in this field is the completion and stabilization of existing standards for open system interconnection (OSI) so that devices from various vendors can communicate through standardised interfaces and protocols.

Since the beginning of 1986 the Esprit project "Communications Network for Manufacturing Applications" (CNMA) has done a lot of progress on this by specifying, implementing, validating and promoting emerging manufacturing communication standards.

To demonstrate the advantages of the principles of OSI within computerizised multi-vendor environments for manufacturing automation using CNMA's OSI communication software and to transfer this technology especially to small and medium enterprises, the "Institut für Steuerungstechnik der Werkzeugmaschinen und Fertigungseinrichtungen" of the University of Stuttgart (ISW) has provided an experimental pilot facility.

2 CNMA's experimental Pilot Facility at the University of Stuttgart

The CNMA pilot at ISW demonstrates the latest OSI communication services. It includes the majority of the project's communication software developments including manufacturing message specification (MMS), file transfer, access and management (FTAM), basic network management (NM) functions and, of major significance, a migration path from proprietary protocols to an OSI environment. The pilot comprises many features of a CIM-environment and provides applications for each of the CNMA's vendors. Computer equipment (minicomputers, PC's and controllers) from each participating vendor is used and by this way the pilot represents a real multi-vendor CIM environment.

2.1 Machine Tool Configuration

To show various aspects of OSI communication within manufacturing applications, the pilot is divided into two independent parts. The first part, which is a fully automated manufacturing cell, is to show time critical data exchange between various controllers for synchronisation and by this to prove and to demonstrate the operability of OSI communication technology under real-time conditions. The second part consists of a single five-axis milling machine and is to show modern communication technology to close the gap in the information flow between computer aided design (CAD), computer aided programming (CAP) and the numerical controller (NC) itself and thus to establish a CAD/NC-link via a LAN.

2.1.1 Manufacturing Cell
The fully automated manufacturing cell is to produce discrete workpieces. It consists of

-a turning centre (INDEX GU 600),
-a machining centre (EX-CELL-O XS 800),
-a local pallet store and
-a linear portal robot.

Both centres are standard numerical controlled machine tools equipped with automatic clamps to fix the workpieces for the machining process. The turning centre has three numerical controlled axis, two linear and one rotary axis. The machining centre has three numerical controlled linear axis and a switched rotating and swinging table.

The pallet store is realized with four conveyor belts arranged in a rectangle were a maximum number of sixteen pallets with one workpiece per pallet can be circulated. For manual exchange of workpieces to and from an external store it has an input and an output position. At a further position the exchange of workpieces to and from the machine tools can take place.

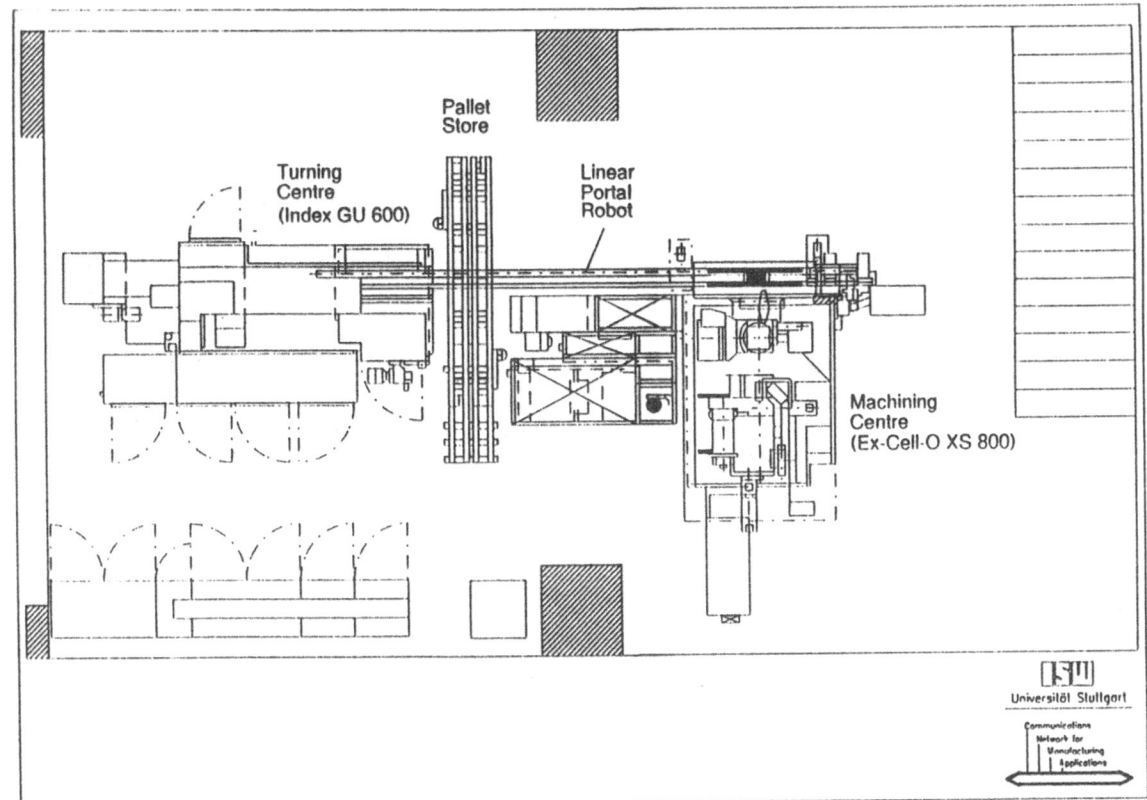

Figure 2.1: Mechanical layout of the manufacturing cell

The workpiece handling between the pallet store and the machine tools within the cell is done by a linear portal robot. It has two linear and one rotary axis, which allow the gripper to move into any position within the portal's workarea.

2.1.2 Five-Axis Milling Machine

The five-axis milling machine (Deckel FP2H) has three linear and two rotary axis. It is used to produce single parts of complex workpieces like models for casting tools. There is no automatic workpiece handling in this part of the pilot. This task, as well as the starting of the machining process, has to be done manualy by an operator. For the machining process the partprograms are downloaded from a fileserver onto the machine controller.

2.2 Computer and Controller Configuration

Fourteen computer and controller devices from seven european vendors are interconnected within the pilot.

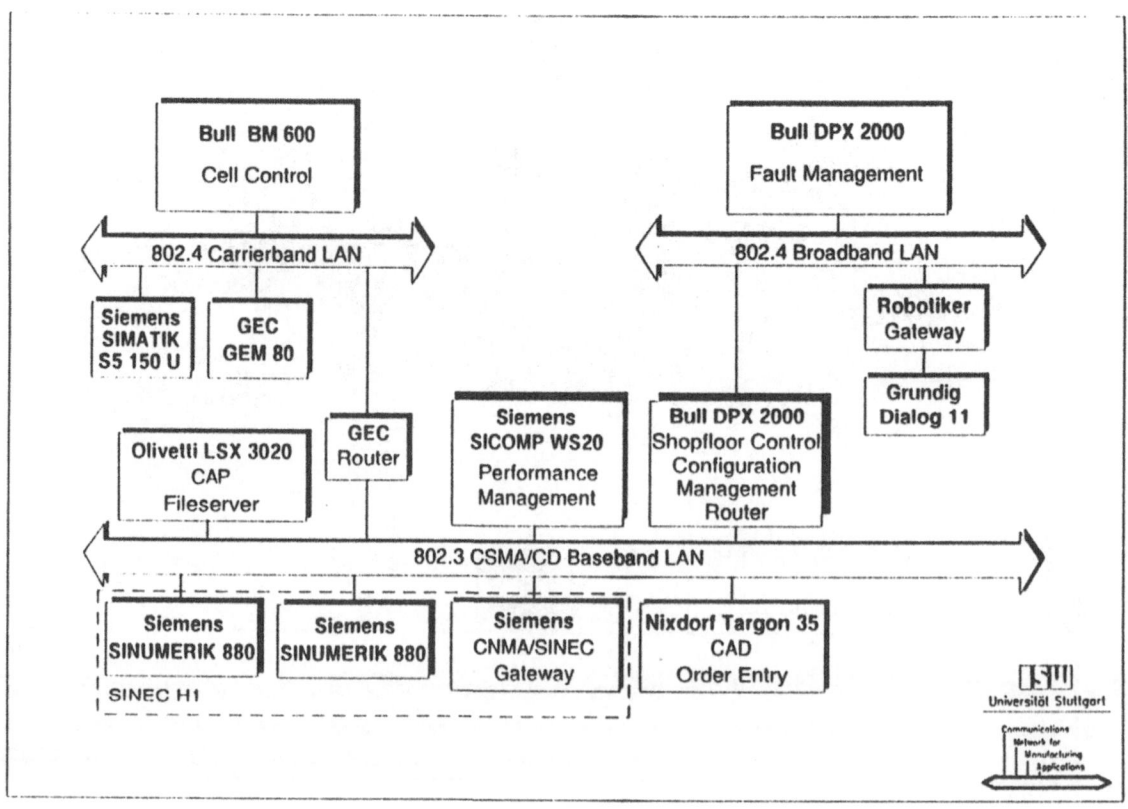

Figure 2.2: Computer and controller configuration

A Nixdorf Targon 35 supports CAD and order entry functions whilst an Olivetti LSX 3020 minicomputer provides CAP and fileserver functions for management of partprograms. A shopfloor control system is running on a Bull DPX 2000 minicomputer as well as a cell control software on a Bull BM 600 personal computer (PC).

Control of the linear portal robot is performed by a GEC GEM 80 programmable logic controller (PLC). The pallet store is controlled by a Siemens SIMATIK S5 150 U PLC. A gateway device from Robotiker supports communication between CNMA protocols and the Grundig Dialog 11 NC controlled five-axis milling machine. Siemens SINUMERIK 880 NCs provide numerical control of the machining and turning centre. These devices use the proprietary Siemens SINEC AP automation protocol. A Siemens gateway device is used to provide communications between the proprietary and the OSI environment demonstrating a migration path.

Systems integration is done by ISW in conjunction with Alcatel-TITN. The software is run over three local area networks (LAN),

-an 802.3 CSMA/CD baseband LAN as specified by TOP,
-an 802.4 broadband LAN as specified by MAP and
-an 802.4 carrierband LAN for low cost MAP implementation.

These three networks are linked by router devices from GEC and Bull.

156

In addition to the MMS and FTAM services used to integrate the application a major implementation of network administration is supported by all the vendors. A fault and configuration management system is provided by Bull and the Fraunhofer Institute whilst a performance management system is provided by Siemens.

2.3 Application Software Structure

The pilot's application software covers six important areas of CIM:

-production management,
-computer aided design,
-computer aided programming,
-shopfloor control,
-cell control and
-machine control.

An overview of which part of the application software is used in which part of the pilot is given in figure 2.3. The shopfloor and cell control software is only used within the manufacturing cell whilst the CAD and CAP software is only used whithin the five-axis milling part of the pilot.

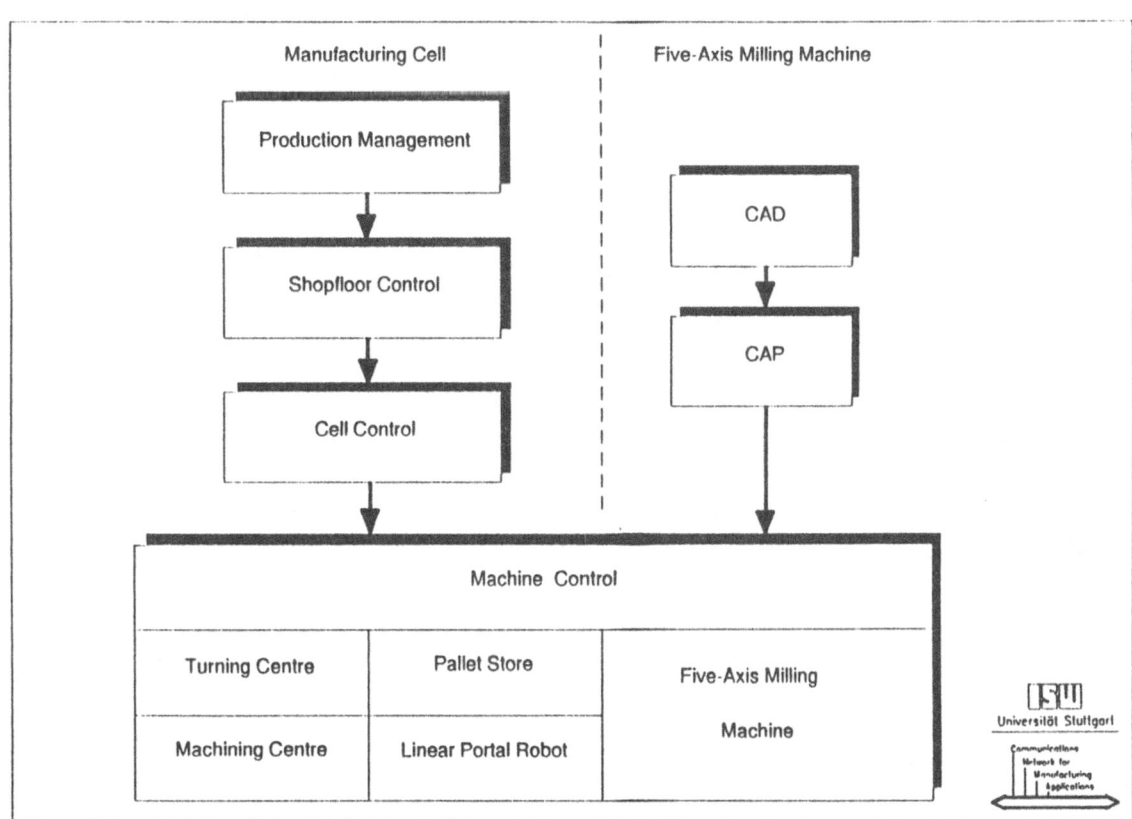

Figure 2.3: Application software structure

2.3.1 Production Management

Instead of a real production management system (PMS) a special development of ISW for simple order entry is used. This software performs solely the elementary functionality of such a comprehensive system which is absolutely necessary for generation, modification and deletion of manufacturing orders to be processed by the shopfloor control system.

Since the ISW pilot does only cover the requirements and the functionality of a single manufacturing cell and not that of a whole factory with many production lines and manufacturing cells and since from the communication aspect this order entry software acts like a real PMS, this simplified solution is sufficient.

2.3.2 Shopfloor Control

For shopfloor control a standard software package called AGIS is used. This package has been developed for Bull and adopted to CNMA's OSI communication facilities by Alcatel-TITN.

Due to the requirements of the pilot configuration the following modules of AGIS have been installed:

-The *"Data Engeneering and Shopfloor Configuration"* allows the configuration of the whole AGIS software package onto a specific shopfloor. It permits the input of a formal shopfloor description (e.g. computer and mechanical equipment configuration) and a formal description of the workpieces which will be manufactured on the shopfloor (e.g. necessary sequence of operations, partprograms, tools etc.).
-The *"Production Management Interface"* receives the orders from the order entry and stores them into the AGIS internal ORACLE database.
-The *"Phase Link"* organizes the sequence of orders for the individual machine controllers and provides a process visualization.
-The *"Manufacturing local Control Subsystem"* controls the machine tools by receiving enabling requests for and finishing messages of the machining process from the cell controller. After approval of enabling requests it sends back positive or negative responses, in case of a positive response together with the respective information for the machining process (number of partprogram etc.).
-The *"Storage local Control Subsystem"* manages and operates the local pallet store.
-The *"Material Handling local Subsystem"* operates the linear portal robot.

All communication with machine controllers, PLCs and NCs, is done by the cell controller.

2.3.3 Cell Control

To reduce the data exchange rate between the shopfloor control system and the relevant machine controllers an additional cell controller was inserted. The cell control software is a pilot specific development by Alcatel-TITN. Due to the internal structure of the shopfloor control software at one side and to the machine controller configuration of the manufacturing cell at the other side, the following architecture for the cell control software has been decided:

-A "*Manufacturing Task*" acts as an agent between the manufacturing local control subsystem of AGIS and the NCs of the turning and the machining centre. Enabling requests to the shopfloor controller are generated if a machine tool is ready for the next machining process. After receipt of a positive response from the shopfloor controller, it causes the download for the respective partprograms and command sequences onto the respective NC and starts the machining process. When the machining process is completed, it causes the appropiate actions to inform AGIS about this event.

-A "*Storage Task*" acts as an agent between the storage local control subsystem of AGIS and the PLC of the pallet store. It causes the respective sequences of actions of the PLC for external and internal input and output of workpieces. In this context "external" means exchange of workpieces between the local and an external store, whilst "internal" means exchange of workpieces between the machine tools and the local pallet store.

-A "*Material Handling Task*" acts as an agent between the material local control subsystem of AGIS and the PLC of the linear portal robot. It causes the respective sequences of actions of the PLC for the transport of workpieces from the pallet store to one of the machine tools and vice versa.

-A "*Relay Station Task*" acts as an agent between the machine controllers within the cell. This has become necessary because all synchronisation between the controllers is done via the LAN instead of using conventional wired I/O-signals.

An overview of the interworking of shopfloor, cell and machine control is given in figure 2.4.

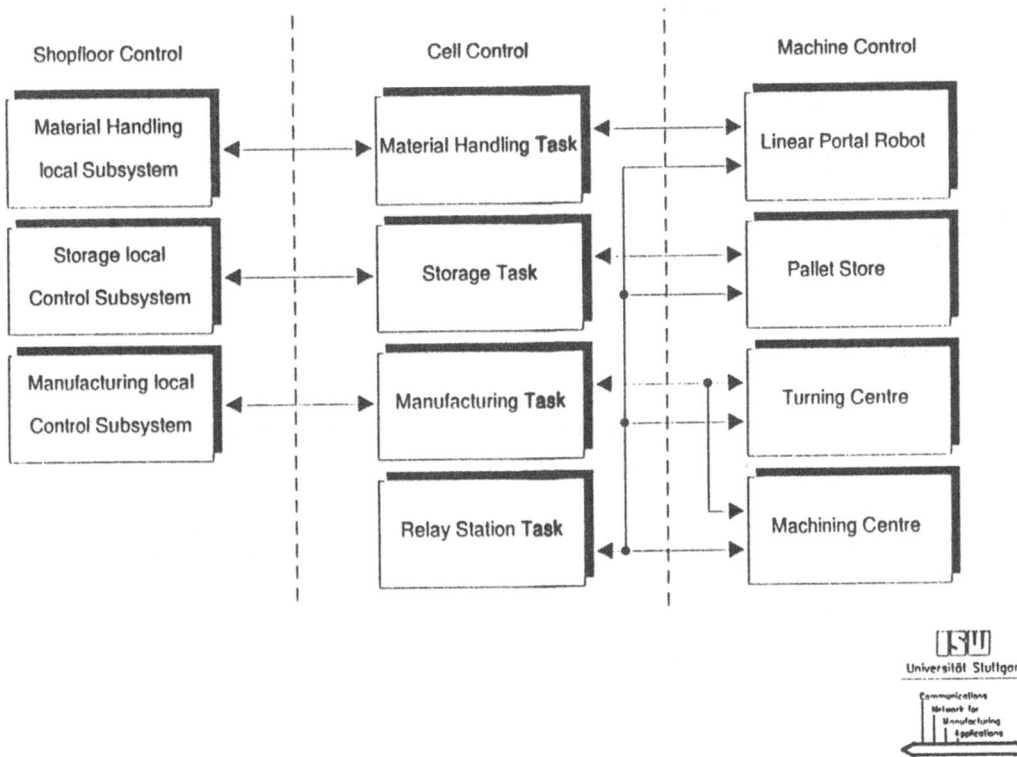

Figure 2.4: Interworking of shopfloor, cell and machine control

2.3.4 Computer aided Design

For computer aided design of workpieces to be produced on the five-axis milling machine the standard software package PROREN is used. This software is commercially distributed by Isycon. It includes modules for two and three dimensional design which allow to design volume oriented models as well as sculptured surfaces for complex surface oriented models. Further modules for data exchange with other CAD or CAP sytems according to the IGES (Initial Graphics Exchange Specification) and the VDA (Verband der Automobilindustrie) file format definition have been integrated.

2.3.5 Computer aided Programming

For computer aided programming a development of ISW called ISW-AX-5 is used. This software is the result of a basic research work of ISW within this area. It converts the geometrical description of a workpiece, which in this case is the output of the PROREN package, into a NC-partprogram for the five-axis milling machine. This is done in the following steps:

-Preparation of the geometrical input data from IGES according to the requirements of multi-axix milling,
-Planning of the machining process by input of technological information like spindle speed, feedrate etc.,
-Plotting of the cutter paths,
-Calculation of the cutter paths including cutter geometry compensation and checking of under-cuttings etc.,
-Transformation of the cutter paths into the maschine's ax-coordinates.

As an additional feature the final cutter paths can be visualized.

2.3.6 Machine Control

To realize the functionality of each machine within the manufacturing cell, special software developments, adaptions, extensions and modifications have been done by ISW for the PLCs of the pallet store and the linear portal robot and for the internal PLCs of the NCs of the two machine tools.

2.4 Network Administration

The pilot's network administration application covers three of the five functional areas of network management:

-"*Performance Management*", which provides the detection of bottlenecks within the network,
-"*Configuration Management*", which provides the administration of communication parameters,
-"*Fault Management*", which provides the detection and diagnosis of faults.

Security and accounting management functions have not been implemented.

Achievement of these functions is done through three different managing processes. The first process is responsible for the workload monitoring, which is a subset of the performance management. The second process provides object management, which is a subset of the configuration management and the third process prepares confidence and diagnostic testing capabilities, which is a subset of the fault management.

In order to integrate the proprietary SINEC network of Siemens into the network administration, the "Institut für Nachrichtenvermittlung und Datenverarbeitung" of the University of Stuttgart (IND) has provided a special network management gateway. This gateway uses the capabilities of the proprietary network management services and adapts them to the services required by the CNMA network administration.

3 Conclusions

The practice from the provision of the pilot verifies the fact that OSI in manufacturing engineering needs introduction step by step. This means that this technology may be first introduced after having done one or more pilot projects with no pressure due to production requirements and after having established the appropriate company internal infrastructure for this technology.

It becomes clear that the vendors have to overtake a leading role during the design and the development of such a manufacturing system. While today's tools or even guidelines for the design, project planning or configuration of such systems are missed, system houses, which have specialized on this market, have to provide consultation, to develop solutions for specific applications and to be master of system integration.

Further more the pure provision of communication facilities between devices of different vendors is not sufficient. Since the vendors have free choice to fix their individual mechanisms for reading or writing data of the device or to give orders to the device, there are still a lot of special solutions necessary in the application software to get a multi-vendor manufacturing system running. These special solutions can first be avoided if companion standards, which define a unified view for automation devices, have been defined and broadly introduced.

Finally, since today the interface between application and communication software is highly vendor dependant, there is still no possibility to design portable application software. This means that today it is not possible to write a really vendor independant piece of application software. Therefore in the future unified application interfaces have to provide a high measure of portability for application software.

Introduction to CNMA
Network Management

P. Martin, E. Arnould

Centre UNIX BULL S.A.
1 rue de Provence
B.P. 208
38432 ECHIROLLES
Tel : 76 39 75 00

Summary

   In today's industries, networks are becoming more and more the backbone of automated systems. The ever increasing complexity of these networks requires powerful tooling to guarantee this efficiency and their maintenability. Network Management aims at satisfying these constraints by integrating all levels of policies for the control of CIM networks. This need is so obvious that it has become a strategic topic of standardization. This paper will introduce CNMA Network Management by providing an answer to the following questions : why do we need Network Management, how much does it save (and cost), what is it for and what is it ?
The first section illustrates the relevance of management in networked computer systems.   We show that it is not only useful to have management ; it will often turn out to be indispensable due to the size, complexity and scale of the problems processed.   The scale of the problem handled is presented in the second section.   Section 3 presents a formal view of the problem space followed by a couple of usage scenarios.   The last section presents the architecture selected to implement the CNMA network management.

1.What is Network Management ?

The proliferation of network resources from various vendors has dramatically increased the networks complexity, creating an acute need for effective management of these resources. As computer networks develop and become standard backbones in computer systems, they reach more and more users. Industrial computer systems are no exception to that trend and network control functions that used to be reserved to computer communication experts must now be handled by all classes of professionals using networked computers in their business. Thus, the computer user's needs have shifted from a need for network resources to a need for networked resources. For most of these people, the network operations remain mysterious. A primary goal of Network Management is to offer as simple a view of the system as possible by hiding all the communication nitty-gritties not relevant to the user's applications to increase the overall system maintenability. Network Management will support the user's understanding of his underlying system, from his point of view, and translate it into precise operations.

Another goal of Network Management is to build distributed computer systems capable of reacting to changing operating conditions : system management must provide self analysis. For instance, the system manager, the top level Network Management application may find it appropriate to activate a machine on the network to share the workload of a process to complete. Likewise, the system manager may find it appropriate to remove a machine from the network due to a solecistic behavior. In such instances, human intervention may be wished to confirm the manager decision or the manager may operate on a totally automated manner. Knowledge based technics may be used to enhance the decision process to react to changing system states.

Network Management helps to reduce the operating costs of distributed computer systems : capable of managing without or with little human intervention, the system manager can react quickly and precisely to changing network conditions. This does not only reduce the operating costs by limiting human intervention, but also ensures a logical and controlled management of the system leading to a better utilization of its resources. Let's consider for instance an intelligent fault management application : when a fault happens, this application is able to diagnose the fault, propose a repair and guide the user in making the repair, leading to a much better mean time to repair (MTTR). Likewise, by observing continually the network, this fault management application is capable of observing degradations leading to a probable failure and therefore to alarm a maintenance operator that a preventive maintenance is likely necessary. Supervision of this type increases the mean time between failures (MTBF). The same observations can be made considering real time performance analysis. The network operator is able to monitor on his console the performances of the system. He can then quickly react to bottlenecks or tune his system to have an optimal use of the system resources.

Network Management brings a very powerful tool : it is dedicated to end users ; yet it is capable of analyzing and reporting all aspects of a system behavior. Based on this analysis, Network Management is able of self correcting its behavior with limited or no human intervention. The development costs of Network Management are high. But the savings it brings to the networked community are tremendous. For each minute of failure saved in a large plant, Network Management buys itself weeks of development time.

Finally, it is worth adding that including Network Management in an existing network or planning it for an upcoming network does not add any constraint in terms of current or future network devices. The Network Management applications are designed to take into account the diversity of existing devices.

2.Scale of the problem handled

Major manufacturing industries like automotive or aerospace industries have expressed a need for distributed computer systems to operate their factory floors. They need a tool that offers integration and intelligence (reactions comparable to that of a human operator) with a very high tolerance for failures and an adequate performance. This is a two sided problem : the technology must be developed to handle this size of problem but we must also ensure a graceful integration of these new techniques among the people that are to use them. This is a challenging field for distributed computer system designers as the complexity of the tasks handled is ever increasing. The goals of the manufacturers using distributed computer systems are clearly set : full scale automatization and control of the factory floor. But the added complexity of the computing environment calls for a better control of the network.

This complexity increase results from three interrelated factors : first, the number of networked machines is increasing ; the reliability of the hardware used in networks allows them to grow steadily. Second, as these networks grow, their performance must increase to compensate for the addition of new machines taking away part of the total network resources. This increase in performance is measured in terms of raw bandwidth increase and network latency decrease or network access time decrease. The bandwidth gives a measure of the amount of information that can transit between two or more machines, depending on the network topology, in a steady state, that is after all the protocol negotiations to access the network have succeeded. The time it takes to negotiate access to the network is the network latency. It is usually measured as the time for two user applications to start exchanging the first data. The third factor for added complexity is the geographical impact due to the addition of new machines. When the area covered by a network is that of a factory floor, problems of fault isolation and maintenance, for example, are complex. The bigger the network, the higher the probability of its failure and the higher the cost of a failure.

As they develop, each of the above three factors takes its toll on the user's understanding and ability to control his distributed computer system.

Distributed computer system designers must not only account for the accrued complexity or failure probability ; they must also map their solutions into a product understandable by, if not familiar to, the human operators. Network Management redefines the role of the user and his computing environment. Network Management offers a wide range of entry points to satisfy all classes of users, from computer system developers to factory operators. Network Management's philosophy focuses on efficiency. It improves the efficiency of a system and of the people using it by offering the right service and level of interface to the right person ; for example, it automatically detects and localizes faulty components, speeding their repair ; it analyzes performances, allowing optimization of the resources used ; it collects information later used for analysis. Eventually, Network Management will integrate the human knowledge of the environment it supports, in an adaptable manner.

3. A random walk through network management

3.1. Management space

The picture below is called the management space. It represents the overall view of a network seen from a Network Management point of view. The lifecycle phases axis depicts the life of a network over time, from planning and designing it to phasing it out. The functions axis depicts the typical management functions that are offered, according to the ISO classification ; in this classification, Network Management services are directed toward the management of configuration, performance, fault, security or accounting. CNMA presently covers only the management of configuration, fault and performance. Configuration management consists of identifying and describing managed objects topology, network components and their relationship. Performance management monitors in real time the network communication quality of service. Fault management identifies and locate faulty components, improves diagnosis and determines corrective actions. This classification does not preclude the existence of services that, for example, reconfigure a system according to performance analysis (configuration + performance management). Finally, the objects axis depicts the management objects. Each family of objects along this axis gives different administrative views or administrative states, that the system must offer to support the management functions.

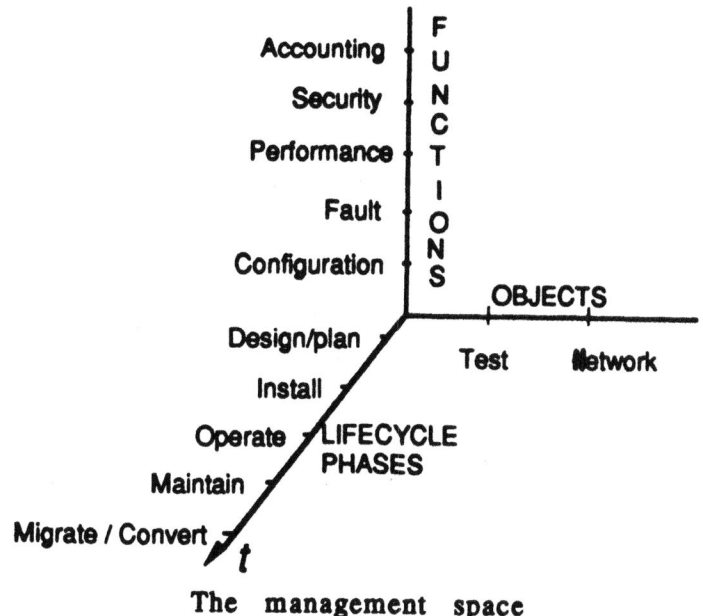

The management space

We count five lifecycle phases : first the design/plan phase. There, the need of a network is evaluated, and solutions are proposed.   Simulations may be done on the various proposals to prove a solution right or to tune parameters of importance to select a choice among others.   At this stage, Network Management must be considered under two aspects :

1 how much Network Management will my solution offer, or how much operational support of my network will be provided by my solution?

2 how can I use my current Network Management system (if I have one) to help me take the best decision?

The planning of a new network, besides its financial cost, is a delicate technical operation. The designer is faced with questions related to expected vs needed performance, desired reliability, overall quality of services, ease of installation and maintenance ...   Selecting a solution that integrates Network Management will prove to be a sound investment. Should the installed network answer only partly the user's needs, the Network Management application will help to pinpoint what improvements must be made. Network Management does not only provide a measurement tool ; it also provides the right services to interpret the results of the measurements.

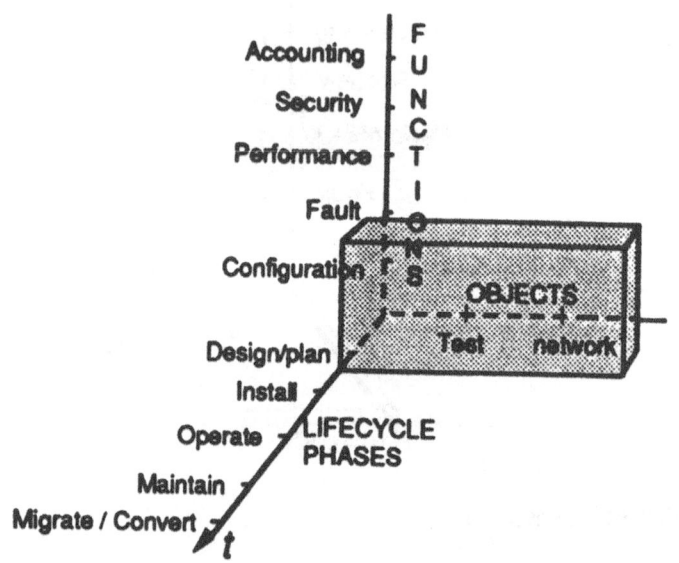

The management space during the design/plan phase

The second stage of a network lifecycle is its installation. Much emphasis must be put on that phase because it may turn out to be much more costly than it first appeared to be. What happens if the network does not work as soon as it is installed ? How much expert manpower will I need to bring the network up ? For how much time ? A network delivered and installed with Network Management has the ability to automatically pinpoint problems and propose solutions or alternatives. It integrates knowledge about how it should normally operate and what may go wrong. At that stage, the network configuration and fault management functions can be extensively used. These functions will depend of the test and network objects. The test objects are used to guarantee conformance to a model and basic communication and interworking of the connected equipments. The network objects provide a functional visibility of the network.

The third and fourth phases usually go hand in hand as far as Network Management is concerned ; the network has been installed and is now in its operational state, from a communication point of view.

All the management functions are useful, though in different contexts. At that stage, only the network objects are used. They are used to diagnose the state of the network, control its performance, add or remove equipments, operate maintenance procedures, provide cost reports and analysis ... For these stages, much emphasis is put on the user interface and the quality of the services offered.

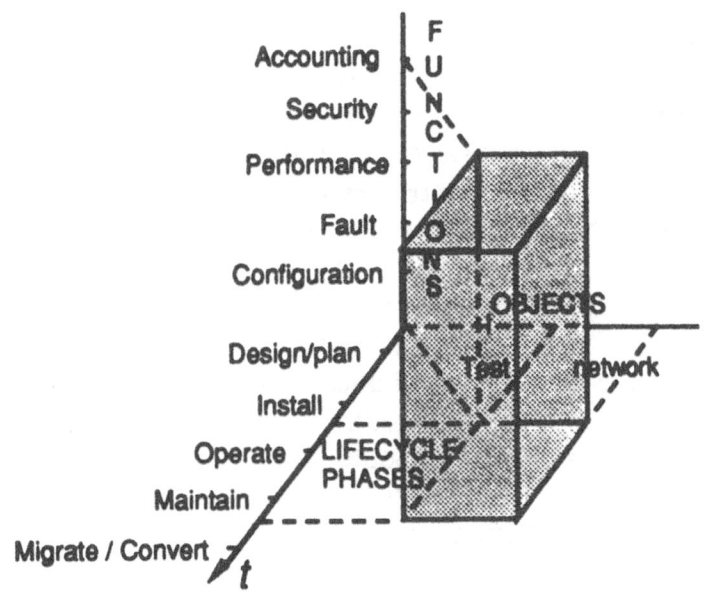

The management spaceduring the operational phase

167

In the last phase, migration, conversion or phase out of the current system, Network Management may again prove of great value by providing maintenance and operational records. Using the tools provided for analyzing these records, the manager will be able to express his needs for a new system.

Currently, CNMA only deals with two types of objects : test objects and network objects. However, its architecture has been designed so that it may easily take into account new types of objects such as application objects (these objects relative to the user applications such as word processing, spreadsheets, ...), operating system objects (file, disk, memory management) ...

The next section presents a few scenarios in which we take a network operator through situations he may encounter. For each scenario, we emphasize the benefits of Network Management.

3.2. Usage scenarios

3.2.1. Installation of a network in a plant

After a planning phase, the network must eventually be installed. This installation begins with a physical layout of the plant, with the operation of cabling. The factory floor, the control rooms and the cable centers are equipped with adequate cabling : it may be twisted pair cable, coaxial cable or optic fiber, depending on the technology of the network that has been chosen. Along with the cables, taps are installed where the network equipment will plug in : programmable logic controllers, numerical controllers, robots, workstations, etc. All network devices that are not visible to the user such as line amplifiers or splitters, are also installed at that time. This cabling operation may be done all in one time or by portions. Testing will be easier in the second case but partial cabling requires the possibility to plug in equipment before the cabling is finished. This is not always possible.

Network Management becomes then a solid testing support partner. The first thing to install is the network management console. This console has a user friendly graphic display that will help the operator in the installation procedure. On this console, the user draws the network : he draws the connected equipment and their connections according to the physical topology of his plant. During this drawing phase, his work is eased by the high level of intelligence of the human interface. It provides him with easy to use commands, understandable messages and plenty of help . The experienced user will find many command short cuts for the frequently used commands. The availability of command scripts opens to him new horizons : a script is a file containing a sequence of commands. It is run on demand, to execute frequently used command sequences. Scripts also receive errors from the network and their behavior may be changed according to incoming errors.

This drawing phase is followed by the logical declaration of the network devices. In this operation, the user binds the devices represented by icons on his screen to instances of real devices available on the installed network. All the available devices and their characteristics and properties are stored in a file. Therefore, the binding operation is normally reduced to the association of an icon on the screen to a device name described in a file. In this case, the system will use the parameters found in the file. The Network Management configuration application then does the rest, as is explained in the next paragraph. It is always possible to add to or remove a device from the file containing the devices description. During this binding operation, the user is permanently helped by the human interface that minimizes the risk of error.

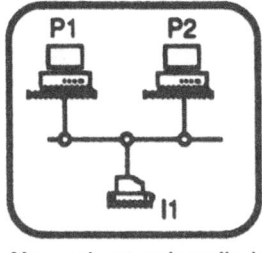 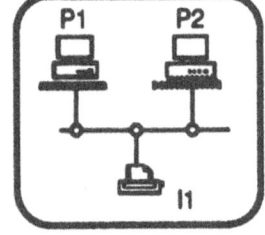

No equipment installed   P1 & I1 installed

Installation   status   display

Following the binding is the installation procedure. During this operation, logical links are set between the Network Management console (this console runs all or part of the network manager application) and the network devices, according to the parameters set during the bind. The devices are declared to the network manager. Then the manager tries to reach each device. If it can't reach a device, it reports an error message to the user explaining why it failed then goes to the next device. Otherwise, if it reaches it, it sets the parameters recorded during the binding. If that fails, an error message is returned to the user. Here again, the human interface plays a major role in displaying the messages and attracting the user's attention to problems arising or status changes by setting the graphical representation with different colors and using sound alarms when needed. Each device is represented on the screen by an icon. By looking at an icon, the user is immediately aware of its status : the device is installed or not (shaded grey or high contrast black and white), the device requires attention (orange) or the device has a major problem (red).

If the application is incapable of pinpointing the exact location of a fault or exactly which device is faulty, it will assume the worst case and report a problem for the suspicious area, therefore changing the status of several devices at the same time. To avoid overloading the screen with useless information, the human interface will usually only warn by an icon status change that the network requires user attention.

The installation operation, executed with the install command, is now over and the network can be monitored and controlled in real time. This operation is part of the configuration application and is used any time a new device is added to the network. The network configuration can be changed at will (dynamic configuration), under control of the Network Management application. There is no restriction to the number of devices that can be contained in a network.

3.2.2. Automatic maintenance procedures

This example goes beyond the area of Network Management since it covers the management of devices outside of the OSI environment. It shows however how easy it is to extend the CNMA Network Management applications.

Most of the devices used in a manufacturing environment require periodic maintenance. Let's now see how it can be automated using network management by doing a case study on a painting robot requiring daily maintenance.

We assume that the robot itself contains some facilities to report its status on request. The normal sequence of operations to maintain this robot is first to disconnect it from the set of machines it is working with. At that time, all the controllers using it must be warned of this configuration change. Following this, a sequence of tests are run to see the operating status of the robot. The result of each test is logged for future use.

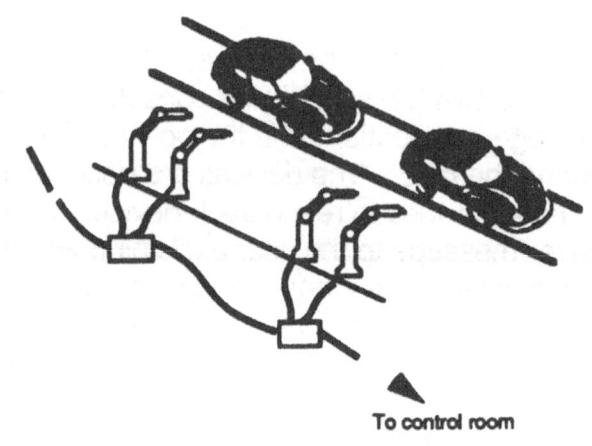

To control room

Manufacturing floor

If all the tests pass, the robot is considered operational. It is reintegrated in its environment and made available anew by warning all parties interested in using it. If any of the tests fails, a maintenance agent must be called to fix all the problems that have been diagnosed. Depending on the gravity of the problem, different persons may have to be called. After the robot is fixed, it is reinstalled in the network as described above.

170

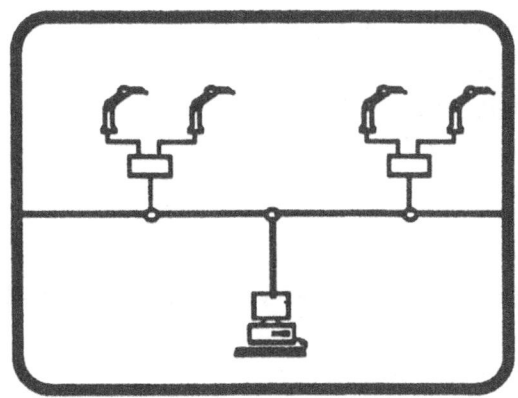

The operator console

Part of the manufacturing floor and the corresponding view on the network administrator console are depicted in the two figures above. In this example, all equipments are doubled to ensure continuity of service. For that matter, the maintenance procedure must react very rapidly to a robot failure while its companion is in maintenance.

An automatic procedure maintenance requires a tool capable of starting administrative tasks at various dates and times and capable of reacting to changing network conditions while the maintenance is going on. Given the number of possibilities that must offer a tool to automate maintenance, we devise the use of command scripts. Command scripts offer a powerful yet simple mean to automate the procedure described above. Here is how it looks :

```
at 12:00 every 24 hours do
on alarm "robot 2 stopped" do
print "emergency repair needed on robot 2; robot 1 is in
maintenance; service stopped on station 1"
with priority high
end do
remove "robot 1" from domain "plant"
if error  none then
print "cannot remove robot 1 from the domain plant"
break
end if
install "robot 1" in domain "test" at level 3
if error  none then
print "test of robot 1 failed"
break
else
print "maintenance done"
end if
```

remove "robot 1" **from domain** "test"
install "robot 1" **in domain** "plant"
if error "robot 1 installed" **then**
print "Could not reinstall robot 1" **with priority** high
end if
end do

The first statement specifies at what time to start the operation and its periodicity. The following statement, between "on alarm" and "end do" specifies what to do if robot 2 fails while we are working on robot 1. It tells the machine to print an alarm message with the highest priority. Next is the actual maintenance procedure. Robot 1 is successively removed from its current domain, put in a test domain where all its functions are tested (install at level 3 is called confidence testing) and replaced in the working domain. Error tests are conducted all along these operations to make sure the procedure is going all right.

4. CNMA Network management architecture

4.1. Openness of the architecture

The CNMA Network management architecture is open : applications must remain opened to future extensions to answer upcoming user requirements. They must conform to the spirit of OSI. This openness is largely supported by the definition of interfaces (API : Application Programming Interface) between the main parts of the architecture. These interfaces between the different architectural levels adhere to the following principles :

Provide interface at points where the service description and number of interactions can be made small

Group related functions

Localize functions

Provide boundaries where interfaces can be standardized

Provide appropriate levels of abstraction

Allow levels to be bypassed

These APIs and the use of standard UNIX environments (Xwindows, MOTIF, ...) allow the development of portable software. Software developed within a given architectural layer sees its surrounding layers through their APIs and is therefore independant of other layers implementation.

Modularity of the Network Management developments is another major point. The modularity of the Network Management is not visible to the user. A module, also called an object, is a self-contained software implementation defined by a

function and an interface.   The services offered by a module are visible only through this interface.   It guarantees the independence among the various modules implementation.
Modularity offers the following advantages :

the various modules can be independently developed

a module can be improved or replaced with no side effects on other modules

the functionality can be increased by adding new modules

the overall system function can be distributed on a network

4.2. CNMA model

The CNMA model is based on the latest developments of the ISO standards for Open System Interconnect (OSI),[4, 3].   It comprises 3 elements :

The management process

Management communication

Managed objects

A management process is an application process participating in Network Management.   At the model level, no distinction is made between an agent and a manager process.   Agent and manager processes are particular instances of the management process applied to specific conditions, with given restrictions. Management communication is a form of communication suitable for the transfer of management information.   A managed object is the administrative view of a resource, i.e. these aspects of resources that are subject to management : on one hand, abstractions of OSI resources such as layer entities, protocol machines, connections and representations of physical resources and on the other hand, abstractions of all other non OSI resources.   A managed object is defined by the attributes visible at its boundary, the management operations that may be applied to it, the behavior it exhibits in response to management operations and the notifications it emits.   The set of managed objects within a system, together with their attributes, constitutes that system's Management Information Base (MIB).

4.3. Architectural components

CNMA Network Management uses the following architectural components [1, 4, 3, 5] :

Management process :
  an application process participating in network
  management. A management process has three
  functional levels as depicted below :

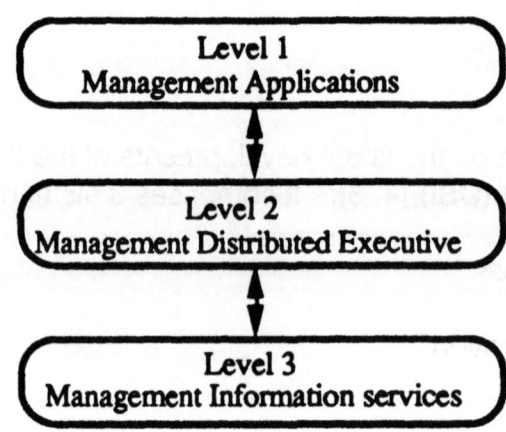

The management process

- Administrator :
  An administrator is a human being the service is addressed to. It is typically a
  human operator controlling or observing the functions of the Network
  Management. The administrator is the primary user of Network Management
  but not necessarily the only one. Some network management services can be
  totally automated and may not require any human intervention.

- Management application :
  a management application is that set of operations that implements a service
  or set of services objective of a management domain.

- MIB (Management Information Base) :
  the management information base represents the set of all information relative
  to the managed system. It is not restricted to a particular management domain.

- Manager process :
  a manager process is a management process which has responsibility for one
  or more management activities. Only manager processes implement
  management process level 1 services (management applications).

- Agent process :
    an agent process is a management process part of a distributed application which resides on each system. An agent process, at the request of a manager process, maintains a set of managed objects. It performs the requested management operations on this set of managed objects and forwards information (notifications) generated by the managed objects to the manager process.

- CMIS (Common Management Information Service)

- CMIP (Common Management Information Protocol) :
    the CMIP is the common communication protocol that supports the manager/agent data exchanges.

- Managed object :
    a managed object is the administrative view of a resource, i.e. these aspects of resources that are subject to management : on one hand, abstractions of OSI resources such as layer entities, protocol machines, connections and representations of physical resources and on the other hand, abstractions of all other non OSI resources. A managed object is defined by the attributes visible at its boundary, the management operations that may be applied to it, the behavior it exhibits in response to management operations and the notifications it emits. The set of managed objects within a system, together with their attributes, constitutes that system's Management Information Base (MIB).

4.4. Specification of CNMA's implementation architecture

The first CNMA implementation architecture consists of one manager and several agents. The manager is what the user sees and his means of access to the management information or the result of its processing. The management information is under control of the agents. Each agent is responsible of a set of objects and of the access to the information contained in these objects: The overall information contained in these objects forms the MIB.

The management process architecture is decomposed into 3 levels [6] :

Level 1 contains the management applications. In the ISO documentation, management applications are classed by functional area. However, this classification is for the purpose of convenience, and in performing management activities, management applications may combine sets of management functions to effect a particular management policy. The functional areas of ISO are : configuration management, performance management, fault management, security management and accounting management. We have added user interface as an extra functional area. The philosophy of CNMA's management applications is geared toward interdependance of the management areas.

To that purpose, each application is designed around the concept of applicative services confined to a given functional area that may be used by any application.

These services are constructed with an object oriented method. They are only visible through an interface. This way, it is possible to efficiently distribute them when and if needed. An example of an applicative service of the configuration management application is the "install" service : it performs logical and/or physical installation of resources on the managed network. This service may be used by any other application. For instance, in the first implementation, the user interface application is using the install service to install network resources selected by a human operator.

Level 2 is the management distributed executive (MDX). It contains :

Environment services
Application support services
Distribution support services

The services offered at level 2 are not necessarily yet subject to ISO standardization. Level 2 also manages the MDB, Management Data Base. The MDB contains all administrative information not contained in the MIB. It is not required that the information representation in the MDB follows the OSI managed objects model.

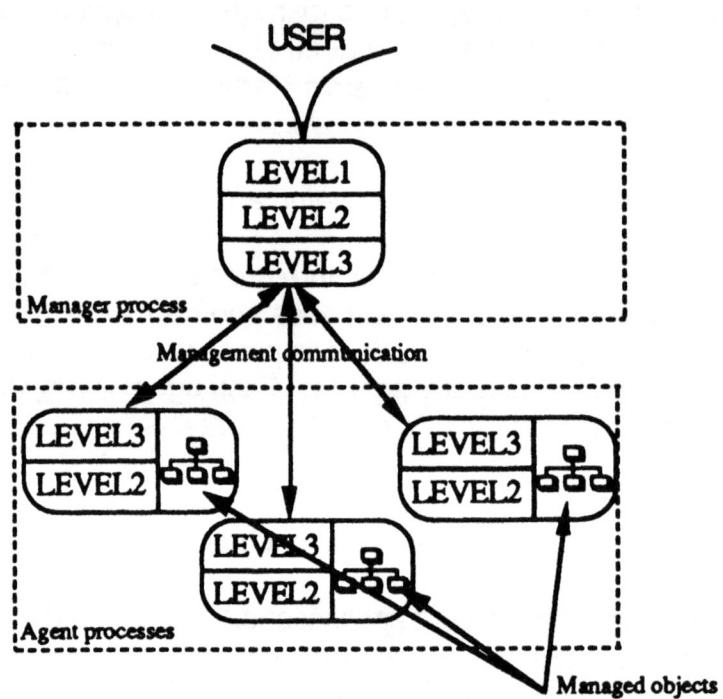

Manager / Agent relationship

The MDB may contain images of MIB objects. Typically, it contains consolidated information obtained by processing the raw MIB information. Information specified for storage in the MDB may become part of the MIB if it becomes a standard managed object. As level 3 services gain in functionality, they may replace then obsolete level 2 services. Any time a level 2 service has a standard OSI management information equivalent service, it becomes a level 3 service.

An example of level 2 service is the GMIB service ; this service gives global MIB (Management Information Base) access : it provides read, write and specific access to managed objects. This module is called "global" because it allows access to several objects and agents with a single request.

Level 3 contains all the basic functions giving MIB access. The MIB represents the set of all managed objects related to a same containment tree, specified in accordance with the GDMO (Guide for the Definition of Managed Objects), standardized de facto or by a standardization body, or specified by a vendor for its customers needs. For instance, level 3 contains the CMIS/CMIP. Level 3 also contains the CL-MIS (Connection Less Management Information Service) ; this module offers the common management information service in a connectionless manner on a per system basis. It provides services to send or receive management information messages that are spread over the network regardless of connection problems.

Each of these 3 levels is accessed through an interface as depicted in the picture below :

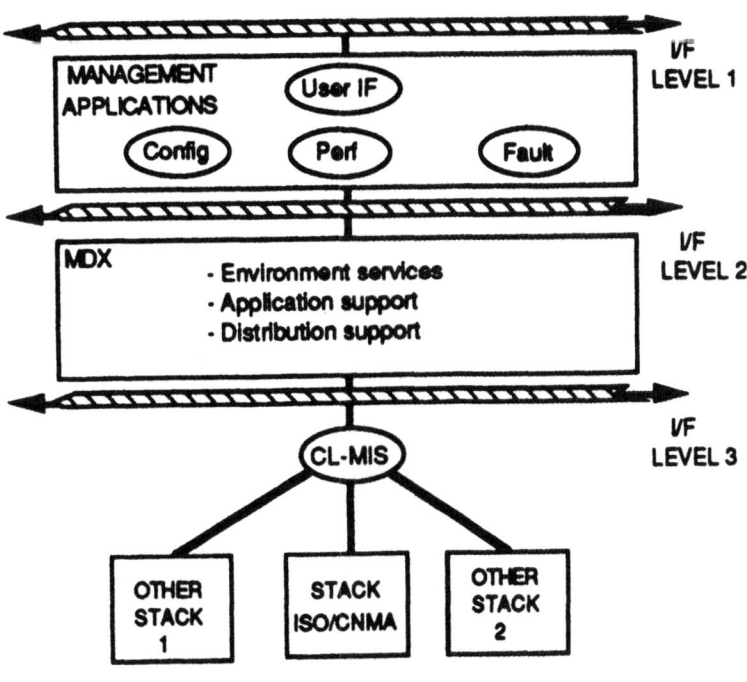

Level 1,2 and 3 interfaces

177

A manager process has all three levels (1 to 3) but an agent process has only levels 2 and 3. The MIB is directly accessed by the agents. Information flows from/to the MIB can be passed to the manager using the level 3 Management Information Services. The following picture depicts a view of the implementation architecture:

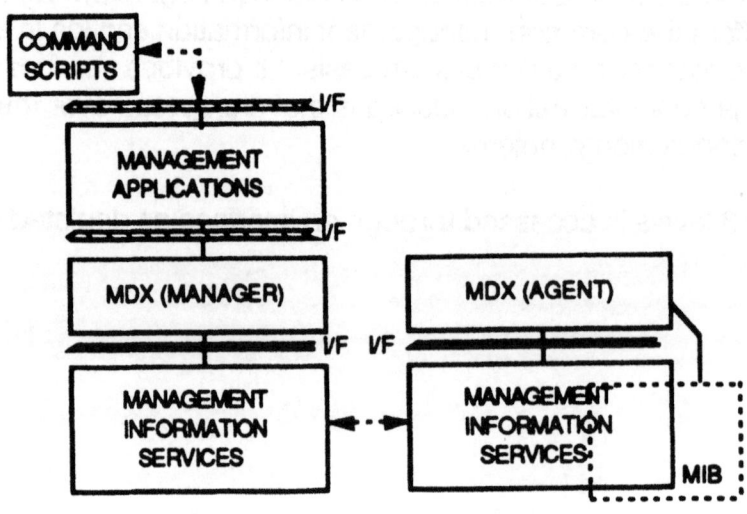

Network management overall architecture

5.Status of implementation

The first two demonstrations of CNMA NMT will be given at ISW (University of Stuttgart, West Germany), and Renault (Paris, France), during the third quarter of 1990. At ISW, three Unix stations performing configuration, performance and fault management manage devices from all the CNMA vendors in conformity with the IG 4.0 and addendum 1. At Renault, a centralized Network Manager running on a single Unix station offers applications supporting configuration, performance and fault management.

Future work on NMT, planned within phase 5 of CNMA, will include management of OSI NM forum objects, the distribution of the manager and security management.

Acknowledgements

The work described in this paper results from the collective effort of all the partners of CNMA and especially Bull's, Fraunhofer Gesellschaft/IITB's, Renault 's and Siemens ' Network Management teams. Particular thanks must go to Bruno Sehabiague and Francois Waeselynck from Bull, Godfried Bonn and Jorg Kippe from IITB, Florence Langlois and Gerard Segarra from Renault and Karl-Heinz Deiretsbacher and Wikhard Kiesel from Siemens.

References

[1] - CNMA Implementation Guide 4.0, Volume 2, A-Profile Part 7: Application Dependent Requirements: Network Management September 1989

[2] - ISO/IEC JTC1/SC21 N 3509,Information Processing Systems - Open Systems Interconnection Management Information Services - SMI Part 4: Guidelines for the Definitions of Managed Objects, April 1989

[3] - ISO/IEC JTC1/SC21, Information Processing Systems - Open Systems Interconnection Management Framework ISO/IEC DIS 7498/4

[4] - ISO/IEC JTC1/SC21 N 3294, Information Processing Systems - Open Systems Interconnection System Management Overview

[5] - Manufacturing Automation Protocol 3.0 Chapter 11: Network Management Requirements Specification, August 1988

[6] - CNMA Implementation Guide 4.0 Addendum 1 :
Network Management Application
February 1990.

IMPLEMENTING OPEN SYSTEMS

Use of Computing Standards in Manufacturing

Gary Blunck

Project Manager
Deere & Company
Deere Tech Services
John Deere Road
Moline, Illinois 61265
USA
Phone: 309-765-5199
FAX: 309-765-4128

Summary

In todays competitive environment, every manufacturing company is striving to improve quality and performance (people and machines) and at the same time to reduce costs. There are numerous examples showing standardization in products and product components and in the machines and equipment used to produce those products. The new frontier is the use of computing standards for the manufacturing. This presentation will address the use of standards in computing hardware, operating systems software, netware, data base management systems, data definitions, user interfaces, data definitions, and application component software.

Thank you for the opportunity to share some examples of activities at John Deere as they relate to the "Use of Computing Standards in Manufacturing". Deere is a 153 year old company with roots in Illinois and has grown to a worldwide company with 22 manufacturing sites. Our growth is a direct result of our product quality and service, which we attribute somewhat to the use of common or standard methods and procedures. Some of these I will be identifying in this presentation, which has a layman's flavor.

The use of standards is not new, but the use of computing standards is relatively new. Industry will typically develop specifications for standards after products have been developed but do not interoperate in a multi-vendor environment. A few examples of the need for standards are: early railroads that had different sizes of rails and different widths of tracks; different sizes of fire hoses and connectors; and electrical outlets. In the railroad example, the rail companies were not able to switch cars between companies. You could argue the competitive advantages and owning a market, but you can also see the advantages of the system that has emerged after the implementation of standards. In the case of the fire hoses, many additional lives and buildings were saved when neighboring fire stations were able to work together and have their equipment interoperate. The third example is in the use of electrical voltages in the U.S. and in Europe. With 110 volts being standard in the U.S. we are able to use computers and household appliances in virtually any outlet, and in Europe they can do the same with the standard use of 220 volts, but we can not interoperate our appliances between the U.S. and Europe. These are good examples that show a common sense approach to the implementation of standards and the benefits to the end users.

Before I get into the computing portion of my presentation, I want to make it clear that industry should be driven to implement standards by the business and functional needs of the company and not by the desire to implement technology for its own sake. Here are some examples of implementations of standards in the product design activities: holes sizes; thread sizes; Metric conversion; most common sizes of hardware: nuts & bolts (I'm sure that all of you can relate to these things within your own homes when you try to replace electrical and plumbing fixtures or when you try to repair appliances). It makes common sense, and is very practical. For international companies, the maintenance, support, and repair or replacement part requirements make the use of standards extremely important.

Here are some examples of implementations of standards in the manufacturing engineering and production areas: tooling; machine tools and controllers (i.e. maintenance advantages); material codes and sizes; electrical, hydraulic, or air (compressed and vacuum) connectors. It makes common sense, and is very practical. You must do these thing in order to reduce your costs and overhead, and be competitive in a world wide market.

In Deere we have been using standards in these activities. Even though your industry may be different than the agriculture, industrial, or consumer product industry, you have similar needs and will achieve similar benefits.

Within computer activities, I think that I could identify 4 types or classes of computing standards:
- Proprietary e.g. DECnet, SNA
- De facto e.g. TCP/IP
- National e.g. American National Standards Institute
- International e.g. International Standards Organization

About any company could build a case for the use of their proprietary standards (e.g. DECnet, SNA, CAD geometry) and those products provide a great deal of functionality within their own environment, and in some cases where other suppliers have written software to interface with those products, we could say that is an example of a multi-vendor environment. However, the disadvantage with proprietary standards is the closed environment that is within that vendor's control.

The de facto standards (e.g. TCP/IP) have emerged by an organization (companies and individuals) and groups of companies working together to produce specifications, but not going through the formal process to get recognition from a standard approving body.

National Standards are well recognized within a country (e.g.- American National Standards Institute (ANSI), National Equipment Manufacturers Association (NEMA)) and may also lead to the acceptance of the standards on an international bases.

International Standards are becoming more familiar within the communication areas via telephone and telegraph applications and are supported by International Standards Organization (ISO) and The International Telegraph and Telephone Consultative Committee (CCITT).

Even though I have identified good reasons for justifying the use of standards, many companies still recognize the competitive advantages for not promoting or implementing standards within their product lines. Companies will argue that they want to protect their markets or that the standards do not have the functionality that is needed or that the standards process takes too long. And I believe that in many cases, the companies are correct; however, we should never-the-less work towards the goals of having open systems and sharing information, which really requires published specifications and adherence to those specifications (i.e. standards).

Examples of standards in computing hardware:
PCs
Unix workstations
Disk drives
Monitors
Printers

Examples of standards in operating system software:
DOS
Unix
POSIX - portable operating system interface developed by IEEE, adopted by NIST, as the initial Federal Information Processing Standard 151, and has been registered as a draft international standard.

Examples of standards in programming languages:
C
FORTRAN
Cobol
Ada
Pascal

Examples of standards in netware:
MAP
TOP
TCP/IP
X.25

MAP and TOP are part of a global effort to standardize inter computer and application networking. This effort come from the need to simplify the task of connecting computing products from different vendors or suppliers to form distributed data processing environments. The Technical and Office Protocol (TOP) program address requirements for the engineering and general office. The Manufacturing Automation Protocol (MAP) Program addresses requirements for the factory floor. Both MAP and TOP share some of the same specifications. Governments of countries have also recognized the needs and benefits of these activities and have formed their own profiles that include specifications from TOP and MAP (for local area networks) as well as other protocols that address Wide Area Networks and international networks. Examples of country profiles are Government Open Systems Interconnect Profile (GOSIP, there is a U.S. version and a U.K. version, and Canadian Open Systems Interconnect Profile (COSIP).

Examples of standards in data base management systems:
Fundamental to any information system is data management. The storage, management and retrieval of data are basic to the integrity and performance of the overall system. Often, diverse business applications have required the use of more than one data base management system (DBMS). Because early DBMS didn't provide the flexibility to serve the needs of different business applications equally well, it was difficult to effectively integrate applications, eliminate data redundancy and provide consistent company wide information management systems. A relational DBMS (RDBMS) helped solve this problem by reducing the data to its simplest elements and then logically linking them in terms of their relationships. A RDBMS can relate data on one data base to data resident on another through a logical linkage of data. Any systems, application or user, can then easily access required data and manipulate it to provide information where needed. This degree of data management, control and flexibility is necessary for a truly open architecture environment.

Examples of standards in data definitions:
Product Data Exchange Specification (PDES) is a data description and format standard under development for the exchange and sharing of all data needed to fully describe a product and its manufacturing processes. Please note the difference between product data and product definition data. Product definition data is the subset of product data that includes only those data elements necessary for the design, analysis, manufacture and test of a product, while product data includes the previous elements and additional elements such as: assembly instructions; process specifications; financial data; customer services information; quality assurance data; testing results; etc. Within John Deere we have our COMMON SYSTEMS that have functionally the same capability that is being addressed by PDES.

Initial Graphics Exchange Specification (IGES) and Product Data Exchange Specification (PDES) are the first components to meet the requirements to exchange product data.

Computer Graphics Metafile (CGM) is the initial component to meet the 2-D business graphics requirements.

Standard Generalized Markup Language (SGML) and Office Document Architecture/Office Document Interchange Format (ODA/ODIF) are the initial components to meet the requirements for document processing.

Examples of standards in user interfaces

FTAM (File Transfer, Access and Management)

MHS (Message Handling System)

MMS (Manufacturing Messaging Specification)

The user interface, X-Windows or the X Windows System makes possible and practical the union of the two most exciting areas in computing: graphics and networking. The Macintosh's graphical user interface launched a revolution. X is a graphically and flexible, multi-tasking, hierarchical windowing systems with build-in links that interface windows to local and remotely networked application. Most importantly, X applications can run on any computer, running almost any operating system, across about any link. There is more to distributed computing than file sharing and printing. Imagine that all the windows on your PC are directly linked to applications, some of which reside on distant mainframes, local minis, workstations and PCs, and each application has the same interface. This is where X and X applications will take you - i.e. linking windows to applications, where ever they may be. To use X, it's more like a programming language than a shrink-wrapped application. So it's not, in and itself, an end user, interactive kind of product. Writing an X application, one that uses windows to communicate with users on a variety of connection, operating system and hardware platforms, is an exercise in application development. It's not a job for the uninitiated. X, at its current level, is more of a developer's toolbox. X applications are what users deal with, not X. There are cost advantages ($2000 to $7000 for an X terminal vs the $5000 to $15,000 for a PC to do the same things). It works like an operating system. You don't really see it, but you know when it's not there. Are X terminals a passing fad? The price is right and they are specifically designed to do their function.

X terminals (terminals or PCs emulating an X terminal) have a screen, a keyboard and mouse, a processor and some memory, and they are running one task, a dedicated application that happens to own the screen while other applications while other applications talk to it.

X has been centered in the Unix environment but programming interfaces are designed to be operating systems independent. DEC sells X on its VMS line, IBM sells it on its VM systems, and could run it in the OS/2 environment. X doesn't run in the DOS environment because it requires a multi-tasking operating system.

There are various implementations of X Windows, such as the Open Software Foundation's Motif and Unix International's Open/Look. They have a different look and feel but are based on the same foundation, and nothing prevents you from running both of these.

Examples of standards in application component software:

Application Component Software:
- Machining standards
- Labor standards (e.g. MIL-STD 1567A)
Techniques - Object oriented programming and data
- Tool kits

Example of an Application Built with Standards - JD/CAPP

General Architecture - Portable and integratable
Hardware - Product development and product porting
OP SYS - UNIX
Languages - C, FORTRAN
Netware - TCP/IP; TOP; MAP
RDBMS
Data Definitions - IGES; PDES; scanning
User Interface - X Windows
Application Component Software

Examples of organizations working with standards:
ANSI
IEEE
ISO
POSIX
X-OPEN
COS
NIST
MAP/TOP User Group
CCITT
JTC

Conclusions

I have tried to avoid any strong endorsement of specific standards, but I wanted to show some of the categories of standards used in computing and some of the alternatives today and the migration towards Open Systems in the future. I am sure that I have missed some of the standards and issues that you have professional interest, but I hope that I have given you some information that will be helpful in promoting the use of standards within your organization, so that you can achieve the real benefits of the use of standards in computing in manufacturing.

Acknowledgements

I thank all of the Deere & Company personnel for their help in providing me with information for this presentation.

IMPLEMENTING OPEN SYSTEMS
MAGNETI MARELLI PILOT APPLICATION

Giorgio Medici

Magneti Marelli S.p.A.
Innovazione Processi Produttivi e Logistica
Via Adriano 81,
20128 MILANO CRESCENZAGO

Summary

The introduction of Information Technologies in modern production systems is a consequence of the large amount of information that must be managed to gain an adequate control of an automated plant.

The adoption of a good standard for all the communications in factory and the definition of the correct strategy to implement it in industrial applications can give sensitive economic and technical advantages.

Inside CNMA project the Magneti Marelli pilot aims to investigate those two aspects providing an industrial application where a common communication technology interconnects various, multivendor devices and computers by a three level control architecture.

The main choices made to integrate the complex environment are:

- make all the differences which can be found at the shop floor level invisible from the cell control level up.

- prefer SW gateways linked to the application software instead of integrating different physical networks by means of specific HW gateways.

1. Introduction

The correct choice of the communication system used to interconnect all the modules of an industrial information system must take two main aspects into account:

- define the correct ratio for cost and performances at each level of the communication architecture inside a plant;

- find a common standard for all the possible connections, looking for the most flexible solution in terms of minimising the future changes needed to upgrade the system when the migration from local proprietary solutions to the assumed standard become possible.

CNMA project has been working for four years on the industrial LANs topic producing an almost complete set of protocols for the communications between

PLCs, NCs, Robot Controllers, PCs and Mini Computers. In the present phase Magneti Marelli has designed a pilot applications which can provide a good test bed of CNMA performances and a significant opportunity to verify some implementation strategy chosen to address the two aspects mentioned above.

Magneti Marelli is participating in ESPRIT-CNMA phase 4 in order to:

- provide user requirements for data communication standards to be used in a real industrial environment.

- develop an integrated information system to manage a production line by monitoring production means and batch tracking, by automated diagnostics and by controlling all the other kinds of messages and commands which are necessary for automated manufacturing. This information system will test the functionalities of CNMA technologies.

- verify advantages and problems arising from the integration of existing control devices and proprietary LANs in a CNMA environment.

The pilot line which Magneti Marelli is going to implement by the end of 1990 for CNMA purposes is a real, actually working production line where car alternators are produced. Different control devices from different vendors are already working on the line, and by CNMA SW/HW Magneti Marelli will connect them together in order to demonstrate the interworking capability of these data communication media.

2. Magneti Marelli pilot description

The Magneti Marelli pilot in CNMA project phase 4 will be implemented at the San Salvo (Chieti) facility. This plant, part of the Electromechanical Group of Magneti Marelli, is about 700 km south of Milan. It is a highly productive and flexible manufacturing unit producing starting motors, rotators for wipers, alternators and batteries for cars and other motor vehicles.

Pilot has been installed on the final section of the alternators production line. This is composed of three different sections on end: shaft production, rotor production and final assembling. Stators are manufactured in another line of the factory and assembled to the rotors in the final section of the alternators line. Nineteen different kinds of alternators can be manufactured at a very high level productivity.

The Assembly Line

Rotors are automatically supplied from the Rotor Line. Stators are gathered in boxes, coming from the Stator Line. The line is composed of 24 stations, divided into two different sections; multivendor control logic and different kinds of control are implemented in this high productivity, high flexibility line (relay logic, programmable logic, pneumatic control, robotised cell, CNC machine tool).

191

The three most significant stations are the two robotised assembly cells and the final test stations.

First robotized cell: an Olivetti Sigma 8600 mini computer controls four arms (13 axes). Every arm can declare the piece to be rejected. All screwing operations include torque control.

Second robotized cell: one Olivetti Mini Computer controls three arms. After any operation the piece can be declared to be rejected and, in this case, subsequent operations are inhibited. If type of alternators changes, after loading the correct part program, the first two arms move to the related conveyor or shaker.

Automatic test stands: two parallel test stands for checking alternators performances (voltage, noise and so on). The handling system is controlled by a Siemens PLC . Two dedicated CPUs monitor the testing procedure and processing data of the two stands. On the main video one can see or define:

- list of types
- parameters per type
- accepted average values
- statistics

Pieces whose performances do not fit the request are translated to the repair station.

3. Main functions of application software

CNMA technologies will be adopted in the Magneti Marelli pilot as the data communication media used by some production management software packages. Specifically the software packages will be dedicated to these activities:

- monitoring
- production tracking control
- diagnostic signals management

Monitoring:

Production monitoring is the collection (real-time and automated when is possible) of the significant information about means' productivity and behaviour of the means. The data collected by the monitoring system are used as a feed-back for the production planning and control. In the work-shop they are also processed as an aid by the machine operators.

Tracking:

Tracking is another kind of shop floor data collection. It concerns all the data necessary to evaluate and traces the work in progress, according to the production planning parameters. Sometimes tracking is assumed to be one of the activities performed by a monitoring system. We think that it is useful to divide two distinguish kind of information about production:

- monitoring: information about means of production - tracking: information about products manufactured

Diagnostics:

Machine diagnostics collects data directly from the machine in order to point out the technical causes of faults. It can be developed as a supplement to monitoring, but it must provide much more detailed information about the state of machines. If the aim of monitoring is to acknowledge that a machine is out of order, the diagnostics system must be able to detect why it is so.

The three production management SW functions have been differently implemented in a three level control architecture (see figure 1); if we refer to a common CIM level definition (1 = production unit controller, 2 = cell controller, 3 = area controller, 4 = plant controller), in the CNMA pilot, Magneti Marelli will develop typical functions at the 1st and 2nd levels, with some extensions at 3rd level, as far as data aggregations, statistical processing and production programs data entry are concerned.

In the specific case of the pilot the allocation of the HW at the different level we have:

level 1

    Siemens PLC (handling/station control) Microprocessors (Test stations)
    Olivetti OSAI CNC (Robot assembly stations)

level 2

    Siemens PCs (PLC Cell Control)
    Olivetti Mini LSX (Robot Cell Controller)
level 3

    Olivetti Mini LSX (Area Controller)

4. Functional specifications of the communication system

In order to manage the complex production system which Magneti Marelli is going to implement with CNMA products, a two-layers control architecture seems to be the most efficient. As shown in the enclosed diagrams (fig. 1) two kinds of physical link will be used in the pilot:

lower level: an 802.4 Carrierband will connect the PLCs (Siemens) necessary to control all the simple manual/automatic stations. It means that only the two robotized assembly stations and the two final test stations, which are the most difficult to interface, will not use a direct connection to a network.

upper level: an 802.3 Baseband will support the data exchange among the PCs (Siemens) and Mini (Olivetti) devoted to manage subgroups of stations and the area controller (another Mini). A Mini (Olivetti) for back-up activities will be connected to the same branch of the LAN.

A bridge from SIEMENS will be inserted between the two branches of the LAN, in order to allow the necessary data exchange among all the devices connected.

Proprietary equipment will be used to connect the relay control interface to the PLC controlling this sort of device, and to connect the conveyor's PLCs to a master PLC as well. This is the configuration adopted to demonstrate the possibility of integrating proprietary communication devices in a CNMA architecture.

In the diagram it is possible to see which device each station will be connected to: the number of the station is indicated within the boxes used to represent the equipment.

Where an operator attends a station, a keyboard terminal with a two LC Display will be connected to the PLC, in order to collect special commands manually and to display alarm or diagnostics messages to the operator.

As far as the communication services are concerned this pilot application aims to emphasise the use of MMS (Manufacturing Messages Services) to manage the communication from the Data Concentrator PLCs up to the Area Manager Minicomputer. The complete CNMA stack with MMS has been implemented on the Area and Cell controllers and on the two PLCs connected to the shop floor devices (fig. 2). All the communications which are performed between the CNMA environment and the proprietary one is managed by gateways (see para 5). The proprietary networks integrated are:

- a SINEC L1, to transmit and receive data from the lower level PLCs
- an OLINET LAN protocol for the connection of OSAI NCs to the upper level baseband network, through an Olivetti PC-250.

4.1. Data acquisition by line devices

In this paragraph the acquisition is explained of the data on monitoring, tracking and diagnostics relating to the line devices, in terms of information to be managed and analysis of information flow.

Monitoring

The basic detail required in order to obtain the scope of the monitoring function, which consists of continuous up-dating of the situation of each device operating on the line, consists of the status of the individual device.

This status shall indicate the condition of the device in as much detail as possible, according to the level of complexity of the device in question. On the line there are relay controlled devices which typically cannot supply very detailed information, while devices exist (e.g. robots) which can provide very important details, specifying with great precision the reason for any production shut-down.

In order to be able to manage even these detailed data easily, it has been decided to structure the status information as follows:

a. The main information, referred to as 'status', consists of the status of the machinery;

b. A further detail is given by the 'substatus' which identifies in greater detail the reason why the device is in the status defined in point a.;

As well as the status information, information on the processing of parts is important. Each device on the shop floor carries out a different operation on the parts, in a time interval which is a characteristic of the device itself. This time can be divided into three parts:

a) loading time: this is the time between the moment when it is decided to load a new part for the next operation and the moment when the operation itself commences;

b) device time: this is the actual duration of the device's work on the part in order to carry out whatever is required by the program operating on the device;

c) Unloading time: the time between the end of the operation and the moment when the worked part is transferred by the device to the line.

The sum of these three times (or the time elapsed from the beginning of a loading phase and the beginning of the next one) constitutes the cycle time, information which is extremely useful for later calculation of the device's efficiency index. This information should be calculated by the line device and should be made available, suitably coded, to the cell controller. The data regarding the cycle time must not be lost, because in order to carry out reliable processing it is necessary to guarantee that all the values calculated during the working of each individual part have been taken into consideration.

Tracking

Tracking is used to manage the data relating to the progress of production. It is therefore necessary to be able to gather the information on the number of parts worked by each device, identified as valid and reject parts, with the relevant rejection statuses. The hierarchically higher levels have the task of resetting the counters present on the devices, in order to obtain correct cataloguing of the data. The characteristic information for tracking is thus production data.

Diagnostic

A special case of status information consists of diagnostic data. These data cover device malfunction statuses due to internal causes as well as accidental causes. The information linked to this concept is the diagnostic data, which is generated by shop floor devices.

4.2 Cell Controller and Shop Floor

With regard to the Monitoring context, the Cell Controller (C.C.) must periodically test all the devices it controls, in order to obtain information on their statuses. With regard to the PLC and test cells, the C.C. PC returns to the PLC 115 PLC concentrators, to request their status or the status of shop floor devices.

At this first level the transfer of data is thus carried out by the polling method, with intervals of approximately five seconds between one inquiry and the next.

For management of the information on cycle times for each operation carried out the fact must be considered that, as already stated, none of these data must be lost. It has therefore been decided to carry out this transfer with unconfirmed services, of the unsolicited type. Each line device (or line device controller) at the end of the operation on a part sends the information on the loading, device and discharge times to the C.C. In this way it is guaranteed that no data can be lost between the production level and the C.C., discharging the responsibility for management of any data code to the cell controller, rather than to the shop floor devices.

In the same way as for the status information, the production data for tracking are also acquired by the C.C. by means of periodic querying of the line devices, using for this purpose services of the confirmed type (management to polling). The records are taken at intervals of approx. one minute.

For the diagnostic data it is the device itself which, having information to communicate to the C.C., takes the initiative to send messages of the unsolicited type. In this way, as well as guaranteeing that the messages regarding the diagnostics are not lost, it is ensured that the controllers are informed of the alarm status of the device as quickly as possible.

4.3 Area Manager and Cell Controllers

The Mini LSX Area Manager (A.M.) shall constantly control the situation of all the devices forming the line, and also of the three C.C.s directly linked to it with application associations. For this purpose, the C.C.s themselves communicate their status or the status of devices to be controlled by them by means of the unsolicited services.

In fact at the moment when the C.C. receives one of the data described above communicated by a line device it immediately sends this information to the A.M., which manages it independently. This means, for example, that the processing of the data gathered for the preparation of statistics, etc. is the responsibility of the A.M. Therefore between C.C. and A.M. the whole transfer of data regarding information acquired from the production line takes place in "server initiated" mode without the A.M. sending any request for information.

This type of approach has been adopted, both in order to use both types of services (solicited and unsolicited) offered by the MMS protocol, and to produce an automatic real-time backup on the A.M. of all the files present on the three C.C.s, with the further result of increasing the traffic on the network.

5. Implementation of gateways on pilot

The integration of the proprietary environments has been realised allocating all the "translation" activities to the SW modules that perform the communication tasks on the Cell Controllers.

This kind of modules are required to manage the dialogue between the Robot Cell Controller and the two OSAI Robot Controller, and the communication between the two Bertola Test Stations and their PC Cell Controller.

This solution allows the connection of the devices on a single physical medium with the same T-profile (the first four layers of the protocol stack is the same as in the CNMA definition). The application SW uses the communication modules as drivers for two different stacks implemented on the communicating devices. As the interface between these drivers and the application software is MMS-like, a future availability of standard MMS communication software on the devices where proprietary protocols are now implemented will hopefully require only the installation of the new drivers.

Management of the communication between robot cell controller and PC-250 Olivetti

The dialogue takes place between LSX 3010 and PC-250: on the latter the "multibrain" application is implemented which already to all intents has the OSAI 8600 robot control function. Physically, the connection between the two units takes place on the baseband line through Olinet proprietary protocol: the cell controller must therefore manage two protocols at the same time, one standard

CNMA and one proprietary (Olivetti). In order to be able to guarantee the possibility of testing the link between the mini LSX and the "multibrain" environment a suitable interactive interface module is provided.

By means of this interface it is possible both to carry out polling procedures permitting gathering of tracking information, and to access diagnostic and monitoring data by means of unsolicited procedures. At the level of the LSX cell controller, status information is also launched in the form of a polling procedure making it possible to test further the status of the device. As the dialogue between LSX and PC-250 takes place by means of a proprietary protocol called specifically to an existing unstandardized environment between the two units under examination there is no adaptability to the CNMA environment.

Management of communications between cell controller and Bertola test banks

As there is no possibility of standard communication between the test banks and the cell controller (Siemens), it has been necessary to make a point to point connection (serial line RS-232) and a suitable communications protocol. In order to be able to guarantee the possibility of testing the connection from the Siemens PC to the Bertola environment a suitable user interface module has therefore been produced.

As the connection between the Siemens PC and the Bertola test station is by means of a point to point line, the information on monitoring, tracking and diagnostic reach the cell level as soon as test banks are available on the concentrators: as a result no MMS service is used in this part of the pilot and there is no compatibility with the CNMA environment.

6. Conclusions

The pilot will be working at the end of this year and by then the solutions adopted in the design of this information system will be verified.

From a user point of view the experience gained till now suggests some interesting point:

- A first possible step towards the adoption of international standards for industrial LANs is to choose those products available on the market which already has joined the emerging standards; it is important at least for the lower layers of the protocol stack (1-4). Up to now it is less expensive to change the communication SW than to change the HW components of a network. (The cabling costs in a factory are relevant because of the specific environment and of the distances that often must be covered to join the nodes of the network).

- CNMA is a good reference to understand what will be the future standard for industrial LANs and thus it can be considered to evaluate the market products to make suitable choices, according with the principle previously explained.

- For the same reason the use of SW gateways is justified to make more cost effective the upgrade of a network.

7. Acknowledgements

We are specially grateful to Olivetti 3S and Telettra for the work developed with us on this project; the results of their studies on the communications aspects of the Magneti Marelli pilot have been included in this paper, mainly for para 4 and 5.

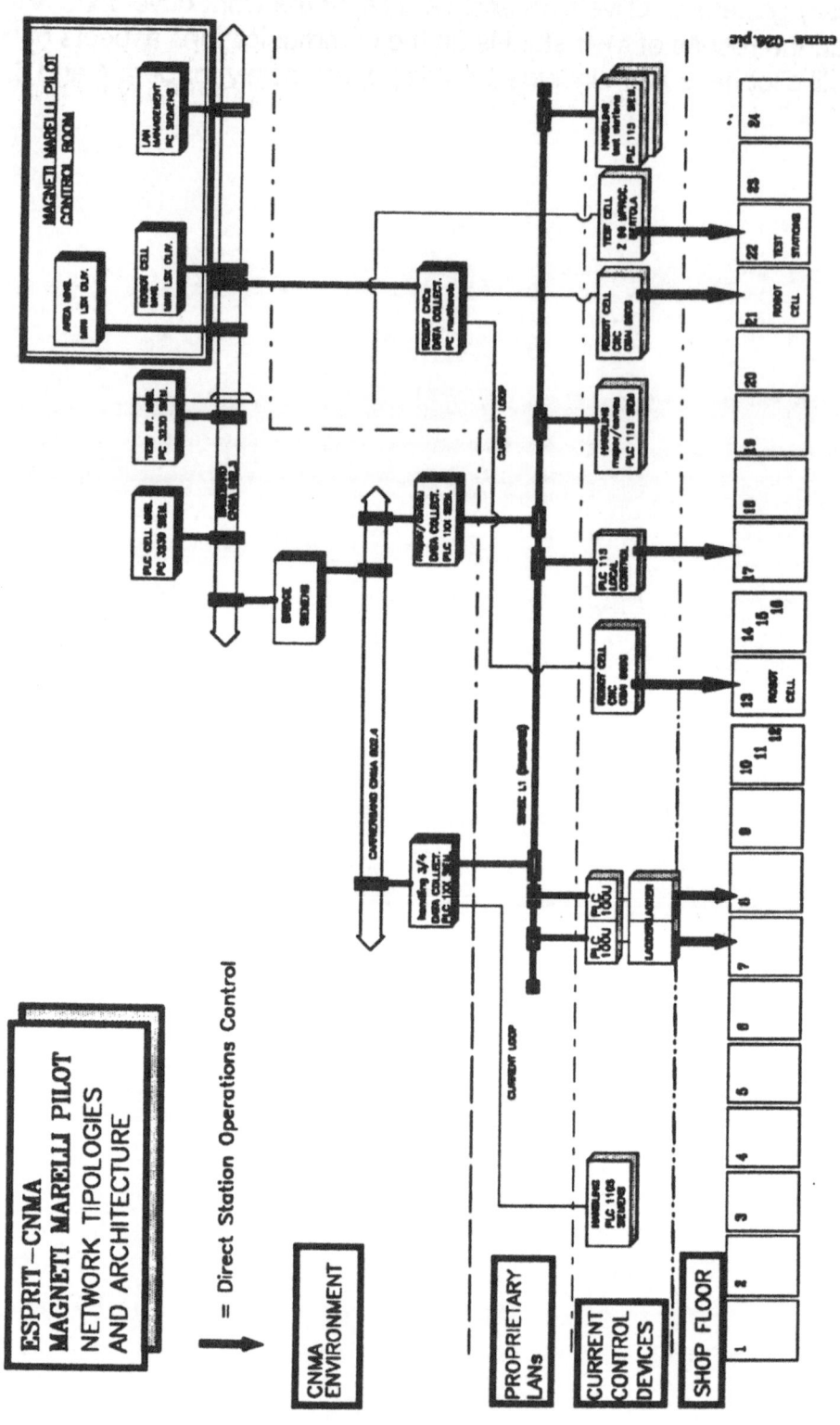

ESPRIT-CNMA
MAGNETI MARELLI PILOT
NETWORK TIPOLOGIES
AND ARCHITECTURE

= Direct Station Operations Control

FIG. 1

200

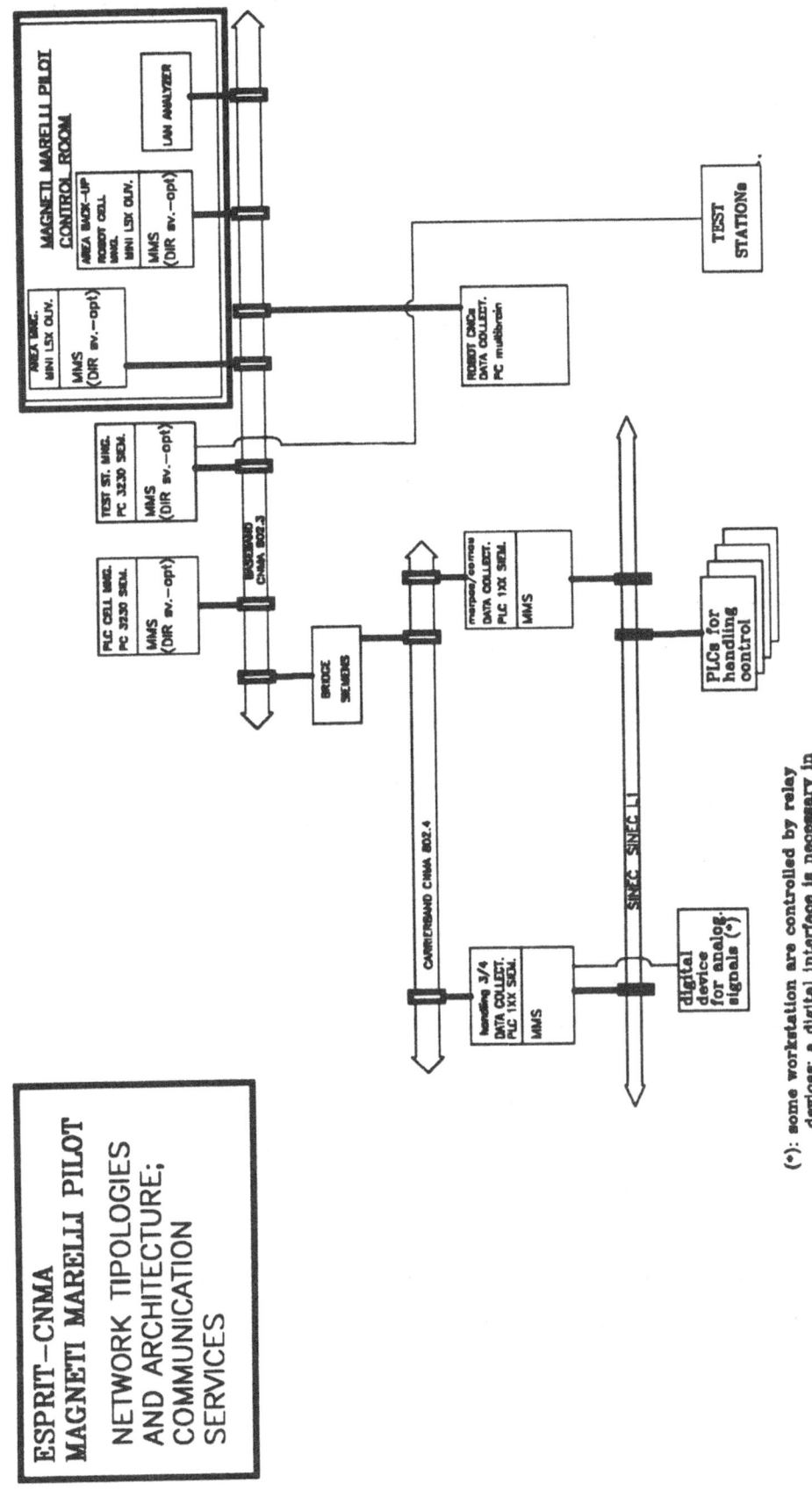

ESPRIT–CNMA
MAGNETI MARELLI PILOT

NETWORK TIPOLOGIES
AND ARCHITECTURE;
COMMUNICATION
SERVICES

(•): some workstation are controlled by relay
devices; a digital interface is necessary in
order to collect the data to be monitored.

FIG. 2

201

ROBOTIKER'S MIGRATION GATEWAY CONCEPT

Juan Manuel Soto
Jesús Sanz

ROBOTIKER
Belako Elkartegia
E-48100 MUNGIA
SPAIN
Phone +34 4 6740002 / 6743139
Fax   +34 4 6743273

Summary

CIM involves many concepts and technologies, but for sure the most important global concept is represented by the word INTEGRATION. Gateways are one of the main integration tools and their use seems to be vital in order to provide a migration path from proprietary protocols to OSI/CNMA.

The number of devices that allow a direct connection to a OSI environment is still limited. This problem is more acute with manufacturing devices than with computers, probably because manufacturing hardware world is less prone to innovate.

Besides that, and even if the number of connectable devices on offer from the vendors increases significantly, there is a big group of Intelligent Industrial Devices already installed and working in the shopfloors all over the world. Although not a simple task, these devices should be made connectable.

The migration period from proprietary protocols to OSI is estimated that migth take any time between 5 and 10 years, depending on many unknown issues such as response from vendors, renewal of the existing manufacturing means, acceptance of standardized communications protocols, etc.

Gateways to OSI LANs provide the solution for the connectivity problem in this period, that it is considered important enough as to give these elements the neccesary relevance.

1. Introduction.

CIM (Computer Integrated Manufacturing) is widely regarded as one of the key solutions that will help to achieve more efficient and better manufacturing than ever.

CIM involves many concepts and technologies, but for sure the most important global concept is represented by the word INTEGRATION.

This concept refers to the capability of having all production means tightly interrelated. The normal way to achieve this is by an intense dialogue between them. There is a clear need for a common communicating nexus to support the neccessary conversational dialogue between these machines, computers and programmable devices (hereinafter called Industrial Intelligent Devices - IIDs).

Only in this way it will be possible to achieve an eficient, unexpensive and reliable way of interfacing the different proccesses that compose a manufacturing plant.

In order to achieve the goal of having a common communicating nexus for all these IIDs, several initiatives have been launched in the last years. Probably for these to succeed they should be based on non-proprietary, international standards, jointly produced by IT vendors and users. This is the case of the ESPRIT-CNMA project.

Some manufacturers of IIDs have already available some of their products featuring a direct interface to these communication systems, besides their own propietary protocols.

Nevertheless, the number of devices that allow a direct connection to this common mean of communication is fairly limited. This problem is more acute with manufacturing devices than with computers, probably because traditionally the manufacturing hardware world is less prone to innovate.

Besides that, and even if the number of connectable devices on offer from the vendors increases significantly, there is an inmense group of IIDs already installed and working in the shopfloors all over the world. Althougt not a simple task, these devices should be made connectable.

Connectability will come by means of retrofitting them with the adequate enhancements. There will be several ways by which this could be achieved :

I - Enclosed in the controller of the industrial device. This could also be :

    a - Provided by the original vendor.
    b - Provided by a third party.

c - Not available.

II - By using an external arrangement.

The solution proposed in this paper is based on option II, because of a number of reasons. Among them :

· Initially, beeing a special add-on, I-a and I-b will surely be expensive options. Besides that, if the machine is to be replaced soon , it might be very likely for such a customized interface to be scrapped with the machine. Henceforth, the position of defending the increasing of investment in such a system is unsustainable. This financial reason will delay the provision of a connectivity path to a great number of "mature" IIDs.

· Provides a solution for case I-c.

· Could provide connection not only to a single IID, but to several of them, thereby lowering the cost per connection significantly.

· It is highly flexible, having the capability of :
- accepting more IIDs.
- being easily adapted to their replacement.
- allowing upgrading of protocols.
- accepting other custom-made software (maintenance, monitorig, etc.)
- having redundant configurations, if desired.
- evolving into a element of distribution production control (cell controller)

· It is based on well-known and easily available hardware and software products.

Within the framewok of CNMA Phase IV, the aim of ROBOTIKER was to provide a cost effective tool that will help the integration into a harmonized production environment of a great number of existing industrial devices. This tool is a programmable network interface unit (PNIU), but for the sake of the simplicity it is usually called a "Gateway".

This introduction must not end without considering the timing of the application of the Gateway in shopfloors. Indeed, because in the long term, when all IIDs will be connection-equiped, the role of the Gateway will surely disminish significantly.

However, what is mentioned here should be considered as a tool that is ment to play a vital role in the migration of current manufacturing environments towards CIM-based ones. This is important when one considers that the migration period might take any time between 5 and 10 years, depending on many unknown issues such as response from vendors, renewal of the existing manufacturing means, acceptance of standardized communication protocols, etc.

This period is considered important enough as to give this element the neccessary relevance within the frame of the present ESPRIT project - CNMA.

2. General description of the ROBOTIKER Migration Gateway.

Since the begining of the project, a proccess of selection was started to find the most suitable Harware and Software basis to implement the Gateway device. When we were considering the requirements that the Migration Gateway had to meet, the following aspects were present :

- · It is due to work in an industrial environment. Therefore, protection must be provided in the following issues : mechanical, electrical, temperature and sealing.
- · It is also a vital element of the chain in the manufacturing proccess. This leads to the provision of the following preventive measures :
  - - Redundant configurations.
  - - Fast replacement and repair : This can be achieved by using a well-known reliable IT vendor as the supplier of the hardware on wich the Gateway will be based.
  - - Modular : In hardware terms, it should be based on a number of cards that plug into a backplane bus. In software terms, modules should be made as independant as possible in order to facilitate upgrading, substitution, addition, and substraction of them with the minimum amount of disturbance at the customer´s site.
- · Finally, it is very important that it should be cost-effective. Therefore, it is needed for it to be
  - - Cheap : based on off-the-shelf hardware and software, except for the gateway-specific software modules. Unfortunately, this excludes ruggerized solutions, which are normally very costly, and heads for a "regular computer in ruggedized cabinet" approach.
  - - Expandable : should be able to cope with a variable number of machines, by means of improving memory and interfaces, which should be done by means of plug-in cards.
  - - High-performance : so that a single gateway could cope with as many industrial devices as possible.

In the view of all these factors, the solution that is likely to be adopted has to employ an off-the-self computer from a well-known and reliable vendor, based on plug-in cards. It has to run a widely known and efficiently supported operating system, which must also be efficient in its implementation for that specific hardware.

For achieving a proper working environment for such an off-the-self machine, a ruggerized, sealed cabinet must be provided for its protection. Other items included in it must be : power supply filtering (and uniterrupted supply, if desired) and the neccessary cooling / ventilation system.

Regarding to the above mentioned aspects and requirements, after a thorough study and preliminary examination and trial of some posibilities, the following choices were selected :

- · **Hardware basis :** PC - 386 machine (BULL Micral 600).
- · **Software environment :** UNIX System V 3.2 from Interactive.
- · **Communication Interfaces boards :**

- CONCORD's MAPware series 1200 for PC Bus System with LLC interface PROMs for the CNMA Network and Carrierband modems.
- RS 232 and RS485 interface boards for proprietary devices.

In terms of basic architecture of the Gateway, we can notice in figure 1, that the Gateway is composed by the following main modules :

· **CNMA stack** that implements the OSI side of the Gateway and performs the function of server in the CNMA Network. It is essential to mention that layers 1 and 2 run in a CONCORD MAPware board, while layers 3 through 7 run in the PC 386 CPU. The upper layers were ported from BULL, one of the CNMA vendors, and interfaced by means of a device driver. The portage also included NMT modules.

. **Proprietary stacks** that implement the proprietary side of the Gateway and are able to act as either Client or Server for the proprietary networks/connections depending on the user-defined configuration. Each stack has also the responsability of providing translation beetween MMS and its corresponding application protocol.

· **Gateway Application**. As observed in figure 1 the Gateway supports multiple connections to proprietary networks. The functionality of the Gateway Application is only to make the message routing between the CNMA stack and Proprietary stacks.

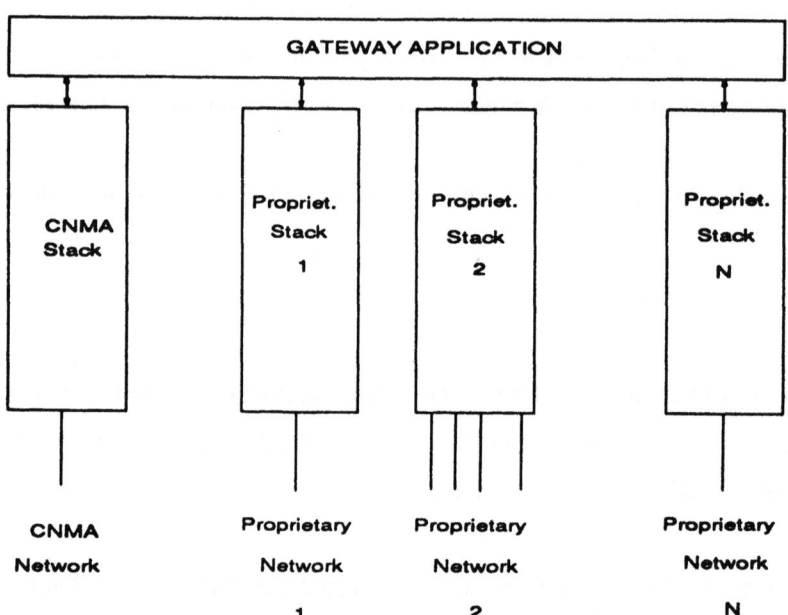

Figure 1 - Basic architecture of the Gateway Device

206

In a more detailed approach to the proprietary stack architecture, it is possible to identify in figure 2 the following modules :

· **MMS Interface Module** that provides to applications a complete or partial MMS Interface, the same that the provided one by the CNMA stack. The special characteristics of this module allows a cell controller application running in the Gateway CPU to make direct use of the proprietary stack. In this way Gateway and Cell Controller functionalities can coexist in the same computer.

· **VMDs Manager.** Each device attached to the Gateway has a VMD in charge of containing a Software image of the behaviour of these devices. This module also implements all the MMS services that have not a direct mapping onto proprietary services.

· **MMS / Proprietary Application Protocol Translator** that is in charge of the bidirectional translation between MMS and Proprietary Application Services.

· **Proprietary Protocol Librairies Module** that implements the complete or partial proprietary protocols.

· **Device Driver** in order to acces to the :

· **Proprietary Network Interface** in which processor could run part of the proprietary protocols.

The rest of the components are two small processes :

· **Timer proccess** used to implement MMS timeouts.

· **Receptor proccess** used to collect incoming messages from the serial board or network interface when this board is not able to interrupt processes running in the Gateway CPU.

3. A review to the implementations made within the framework of CNMA PhIV.

Within the framework of CNMA phase IV, ROBOTIKER has produced three different Gateway implementations that have been installed and evaluated through two pilot installation : the experimental pilot at the University of Stuttgart and the industrial pilot at AEROSPATIALE.

For each gateway implementation, we are going to make a short reviewing of its functionality, connected devices, MMS services supported and special features.

I - CNMA 4.0 / GRUNDIG DIALOG 11 DNC protocol Gateway (Univ. Stuttgart Pilot Installation).

By this implementation we allow that a GRUNDIG DIALOG 11 can interchange information with other devices

207

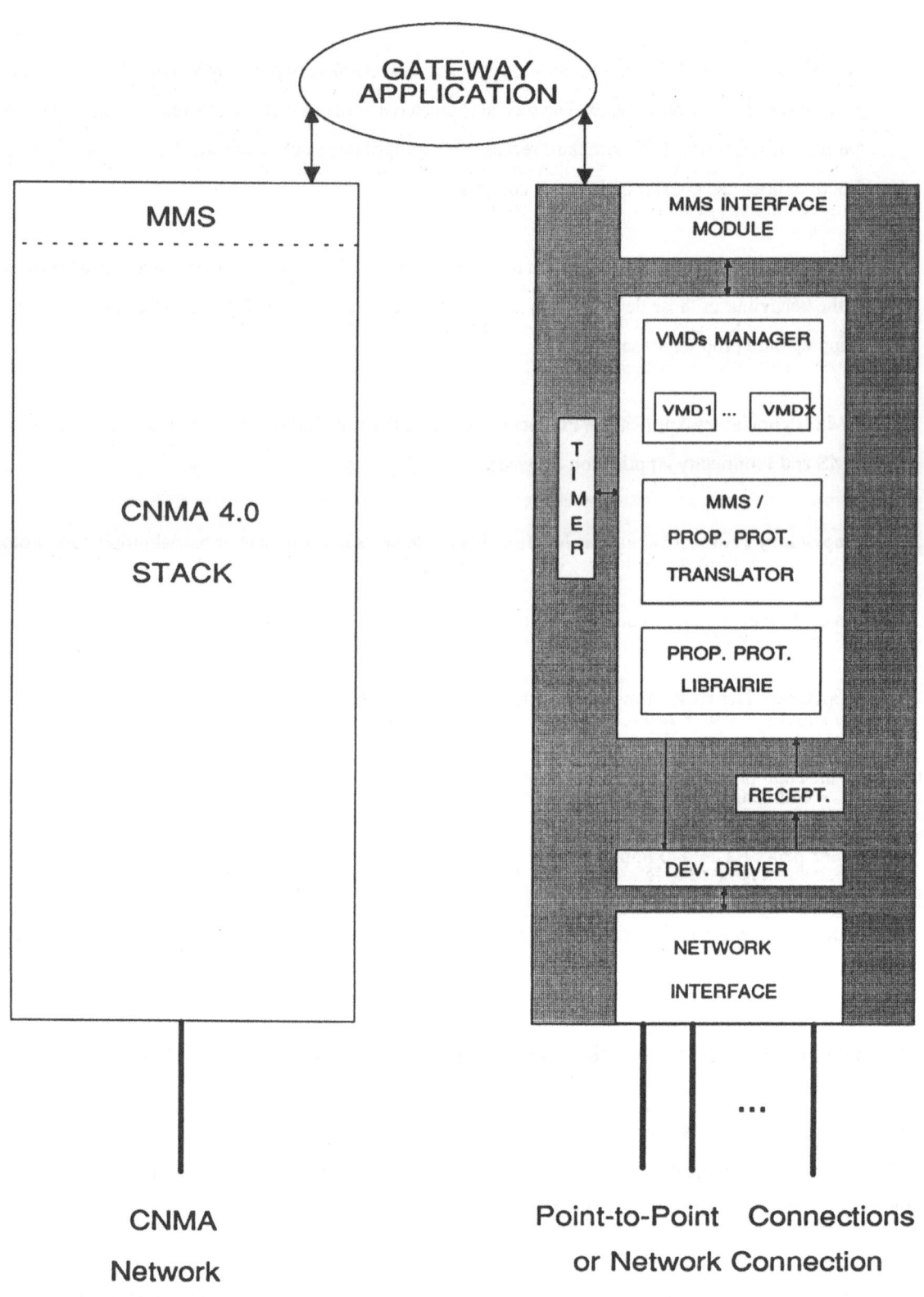

Figure 2 - Architecture of the Proprietary Stacks

through CNMA Networks. This Gateway is placed in a section of the pilot that is used to demonstrate the advantages of LAN communications in the area of CAD/NC link (see figure 3). The geometrical description of a designed workpiece is transferred from a Nixdorf Targon into the Olivetti microcomputer. On the Olivetti microcomputer the NC program generation is performed using this geometrical description. NC programs are then downloaded to the NC using the MMS services provided by the Gateway. This NC controls a 5-axis milling machine where part-program execution is started by an operator. Due to the fact that this machine has not automatic material flow and the low-level features of the communication protocols of the GRUNDIG NC, communication with this device is restricted to download and upload programs initiated by the operator of the NC. For this reason the Gateway only provides for this application the services included in table 3.

In terms of Hardware, there is no a special board to communicate Gateway and NC. All the GRUNDIG DIALOG 11 DNC protocols run in the Gateway CPU and messages are interchanged by means of a standard RS 232 board and

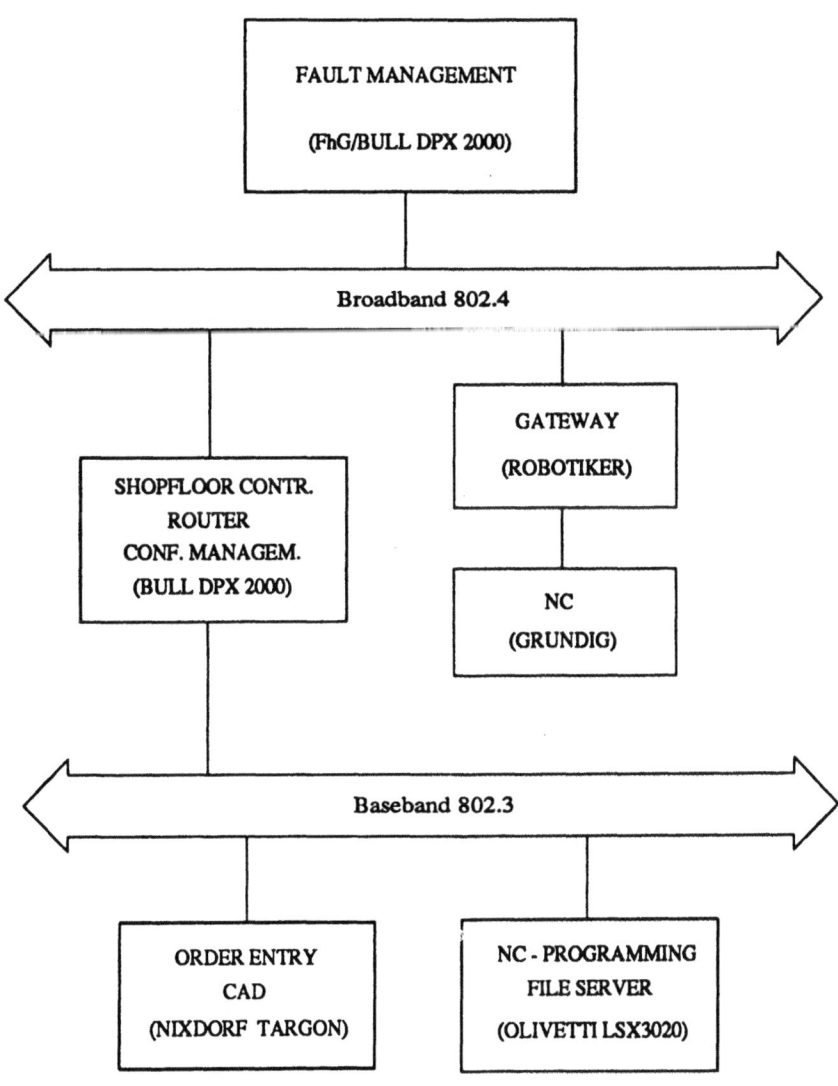

Figure 3 - Topology of U. Stuttgart Pilot

regarding to the OSI side, as observed in figure 3, the Gateway is connected to 802.4 Broadband segment.

II - CNMA 4.0 / UNITELWAY Gateway (AEROSPATIALE Pilot Installation).

This second implementation provides translation between MMS and UNITELWAY messages, allowing that a total of seven NUM 760 Numerical Controllers can interchange information with other intelligent devices through CNMA Networks (see figure 4).

The NCs devices control 4 machining centers and 3 lathes and, by MMS links provided by the Gateways, exchange data with two different types of applications : CMA (Cell Management Application) running in the same computer that the Gateway device and performing cell controller tasks, and MMA (Maintenance Management Application) running in a BULL DPX 2000 that has the responsabilty of collecting maintenance information from all the devices of the shop. Being most part of the NUM 760 unattended NCs, in this installation Gateways provide a complete list of MMS services that can be examined in table 1.

In this case, the UNITELWAY application protocol, UNITE, runs in the Gateway CPU and the low-level protocols run in the processor of a special serial board (ACL-Stargate) with 4 output channels. The Gateway device interchange messages with other CNMA devices througt a 802.4 Carrierband Network.

III - CNMA 4.0 / JBUS Gateway (AEROSPATIALE Pilot Installation).

This last implementation also placed at the AEROSPATIALE site, allows that a series of APRIL PLCs connected by means of a JBUS Network, can interchange information with CNMA devices. These PLCs control the handling and transport system.

The applications that talk with the JBUS Network using the Gateway device are : MMA, as in the second implementation, and SHT (Storage, Handling, Transport Cell Application) that runs in the same CPU that the Gateway device. Table 2 displays the MMS services used for such dialogue.

We have to mention in this case that, the complete set of JBUS protocols runs in a special network interface, the board APPLICOM PC 4000 pluged in the Gateway PC Bus.

4. Conclusions.

Gateways constitute a basic tool for System Integrators because they allow to integrate in the OSI world those industrial devices that have not a direct connection. But, in many cases the impossibility of Migration is a serious obstacle in integration strategies. In this circumstance Gateways also become basic tools since they provide a migration path when users are expecting that vendors incorporate OSI communications to its products. In this way they make feasible transitory solutions of integration.

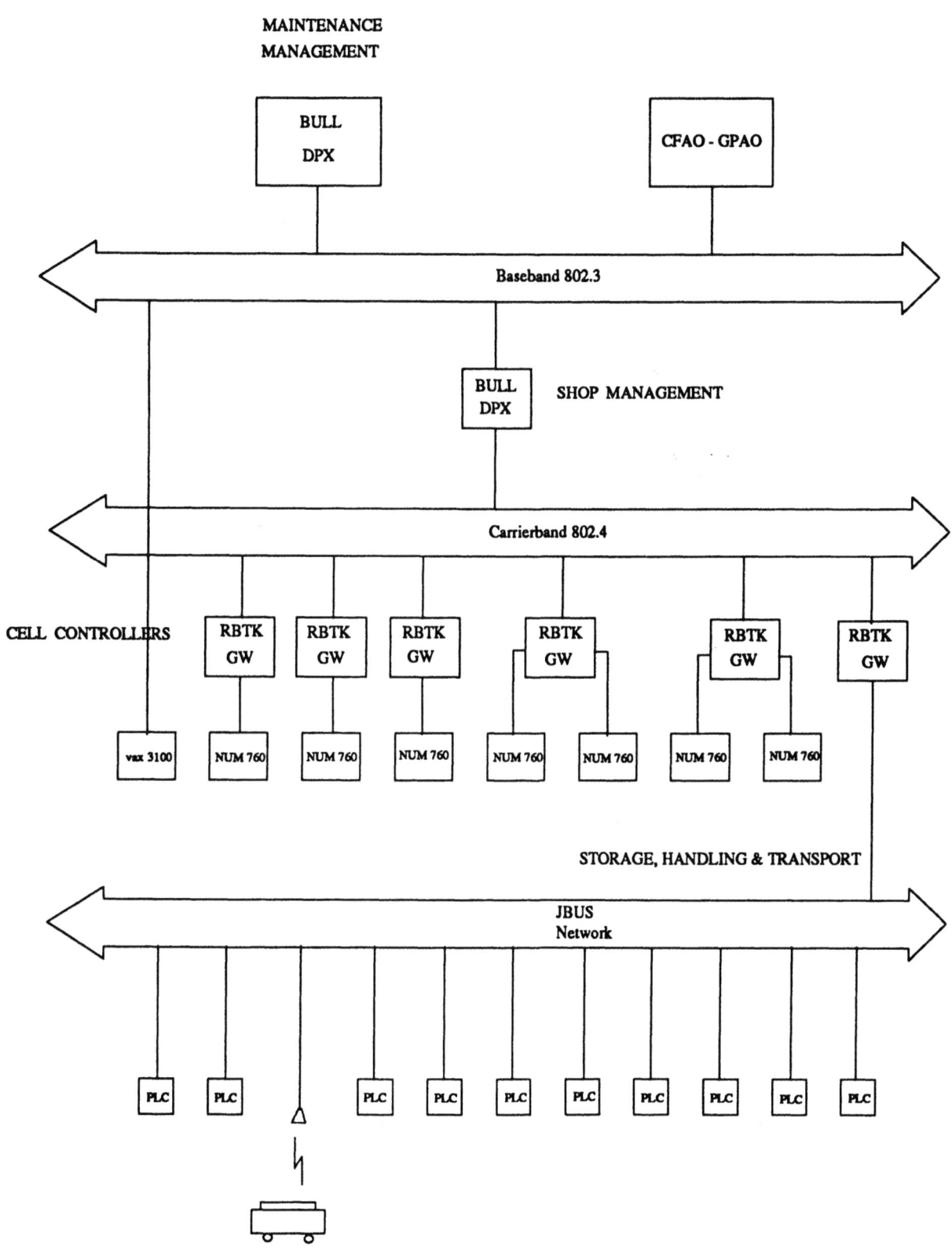

MAINTENANCE
MANAGEMENT

BULL DPX

CFAO - GPAO

Baseband 802.3

BULL DPX   SHOP MANAGEMENT

Carrierband 802.4

CELL CONTROLLERS

RBTK GW   RBTK GW   RBTK GW   RBTK GW   RBTK GW   RBTK GW

vax 3100   NUM 760   NUM 760   NUM 760   NUM 760   NUM 760   NUM 760   NUM 760

STORAGE, HANDLING & TRANSPORT

JBUS Network

PLC   PLC   PLC   PLC   PLC   PLC   PLC   PLC   PLC   PLC

Figure 4 - AEROSPATIALE Pilot Installation

211

CNMA 4.0 - UNITELWAY Gateway

MMS SERVICES

CONTEXT MANAGEMENT
Initiate
Conclude
Abort
Reject

VMD SUPPORT
Status
UnsolicitedStatus
GetNameList
Identify

VARIABLE ACCESS
Read
Write
InformationReport
GetVariableAccessAtributes

EVENT MANAGEMENT
EventNotification

DOMAIN MANAGEMENT
InitiateDownloadSequence
DownloadSegment
TerminateDownloadSequence
InitiateUploadSequence
UploadSegment
TerminateUploadSequence
RequestDomainDownload
RequestDomainUpload
LoadDomain
StoreDomain
DeleteDomain
GetDomainAttributes

OPERATOR MANAGEMENT
Input
Output

PROGRAM INVOCATION MANA
CreateProgramInvocation
DeleteProgramInvocation
Start
Stop
Resume
Reset
GetProgramInvocationAttributes

- Table 1 -

CNMA 4.0 - JBUS Gateway

MMS SERVICES

CONTEXT MANAGEMENT
Initiate
Conclude
Abort
Reject

VMD SUPPORT
Status
UnsolicitedStatus

VARIABLE ACCESS
Read
Write
InformationReport

- Table 2 -

CNMA 4.0 - GRUNDIG D11 Gateway

MMS SERVICES

CONTEXT MANAGEMENT
Initiate
Conclude
Abort
Reject

DOMAIN MANAGEMENT
InitiateDownloadSequence
DownloadSegment
TerminateDownloadSequence
InitiateUploadSequence
UploadSegment
TerminateUploadSequence
RequestDomainDownload
RequestDomainUpload

- Table 3 -

Moreover, it is neccessary to emphasize that Gateways are :

- · Cost effective tools.
- · Flexibles and configurables.
- · Expandables.
- · Transparents.

6. Bibliography.

· "Through MAP to CIM" - Department of Trade and Industry, U.K.

· "AEROSPATIALE Detailed Pilot Functionality Specification. Draft 2.0" - G. Guilbert, AEROSPATIALE, France. April 1990.

· "Experimental Pilot at the University of Stuttgart. Deatiled Functionality Specification" - ISW, Germany. December 1989.

· "Migration Gateway Functional Specification. Draft 2.0" - ROBOTIKER, Departamento de Comunicaiones Industriales, Spain. December 1989.

COMMUNICATIONS FOR MANUFACTURING
OVERVIEW OF CNMA PRODUCTS

T.E. Simmons

BAeCAM
British Aerospace
Guild Centre
Preston

The purpose of this paper is to briefly summarise the relationship between the MAP 3.0 profile and the CNMA IG 4.0 profile.

1. Introduction

One of the CNMA project's aims is to specify a communications system for computers and programmable devices.

Communications in CNMA is based on the Open Systems Interconnection basic reference model (OSI/RM), which is defined by ISO in IS7498. It defines a framework for communications, that is the services to be provided to application processes, the breakdown of the communications software into 7 layers and the split of services between the layers. The model is known as the "OSI seven layer model".

Each service requires protocols - specified interactions - between the two communicating systems. Definitions of the services and protocols in each layer are provided in individual standards documents. However a given service can be provided by a number of different protocol combinations in the lower layers. Hence, an additional document is required to identify the exact protocols used at each layer. Such a selection of protocols is known as a "profile".

2. CNMA and MAP

The CNMA Implementation Guide defines the communications profile chosen for use in the project. It represents a considerable amount of work by the participating companies and its publication is one of the primary functions of the project.

MAP V3.0 is also a communications profile based on the seven layer model. However, when MAP V3.0 was specified protocols for layer 7 were not stable and so interim solutions were included.

The main purpose of CNMA is to focus its research into layer 7 issues, that is the application layer. CNMA has contributed to the MAP evolution through its research and implementation work in this area, so aiding the definition of international standards profiles of communications for use in the manufacturing environment. CNMA has always maintained a close relationship with MAP and other relevant initiatives and also with the standards supported in Europe via such groups as SPAG-CCT, and through the standards bodies such as CEN/CENELEC and EWOS.

3. The lower layers

The CNMA Implementation Guide is the specification document of the CNMA profile. It devotes a chapter to each of the lower layers of the profile, and a chapter to each Application layer service. For layers 1 and 2, CNMA uses Local Area Networks (LAN's) and acknowledges the benefit to users of providing a choice of LAN type. This allows a user to choose a type of LAN based on : cost; performance; installed base; maintainability; or other considerations. Three options are specified. These are 802.3 baseband, 802.4 carrierband and 802.4 broadband.

MAP initially opted only for the broadband technology and has extended this to cover carrierband, with token bus access, but a recent study in Europe has shown that 98% of LAN installations have opted for baseband technology. CNMA's choice of baseband LAN is the same as that supported by TOP.

In CNMA layers 3 to 5 are designed to conform with MAP. These layers are considered to be more stable than layers 6 and 7, and are therefore defined as 'background' for the project. This enables work to be concentrated on the other upper layers.

For layer 6 (the presentation layer), CNMA has specified the use of the kernel functional unit.

4. The application layer

There are a number of layer 7 Application layer protocols specified in the CNMA profile.

CNMA's MMS is based on the IS specification whereas MAP MMS is based on DIS. In the absense of companion standards, the current CNMA Implementation Guide specifies subsets of services for PLCs, NCs, and other devices. Further work is now planned to develop true Companion Standards.

Like MAP, CNMA specifies FTAM for exchanging files between devices.

Much work has gone into the development of Network Management with CNMA. This is a function which permits a remote device on the network, the Network Manager, to access communications related attributes, for example counters or timers in the communication software of other devices onthe network. Using these protocols the Network Manager can monitor and influence network performance, re-configure the network, or perhaps diagnose fault conditions. This management is achieved by writing to and interrogating the management information base on the agents. This is currently the greatest area of activity within the CNMA project, and as a result CNMA now specifies many more managed attributes than the original MAP V3.0, allowing much more sophisticated control of the network by a Network Manager.

Another Application layer protocol is Directory Services, which supports storage and interrogation of information about network related objects in order to provide services such as network "white pages" and "yellow pages" searches.

Also specified in the CNMA Application layer are standardised interfaces for MMS and FTAM. These are to improve application software portability.

5. Products of CNMA

CNMA is a pre-competitive development project. However, many of the vendor partners within the project are now releasing products which have been developed from the software produced within CNMA.

These products are not being marketed as CNMA products. Any attempt to do so would simply cause confusion in the communications market. Compatibility with MAP is one of the aims of CNMA, and these CNMA based products are MAP V3.0 compatible. However they represent an advance on MAP V3.0 because they have been enhanced by the inclusion of advanced ISO/OSI features.

Consequently the products produced from CNMA software are marketed as either MAP V3.0 products, or as MAP V3.0 with OSI enhancements. This latter category are nevertheless compatible with MAP V3.0.

6. Conclusion

In conclusion then, the CNMA profile is a useful vehicle for developing European expertise in the use of MAP V3.0, and in International Standard communication techniques beyond MAP V3.0.

The CNMA vendors are using the CNMA experience to help them to develop MAP V3.0 products which are compatible with the latest OSI standards, as well as with MAP V3.0.

CONFORMANCE TESTING

Kym Watson

Fraunhofer Institute
of Information and Data Processing (IITB)
Karlsruhe, Germany

Summary

Conformance test, that is the test that implementations comply with given specifications, is a wide area of growing significance wherever standards and multi-supplier environments are becoming more common place. In this paper we give an overview of conformance testing activities which have either taken place within the ESPRIT Project CNMA or have resulted from such activities. In particular, conformance testing for CNMA Phase 4 at the ISW pilot site, conformance test for the OSI Network Management Forum, and the marketing of the test tools are covered. Finally, accreditation and certification issues are dealt with.

1. Introduction

The ESPRIT project CNMA (Communications Network for Manufacturing Applications) recognized from the beginning that conformance testing is required to verify that vendor inplementations match the given specifications and that this is a first, major step towards successful interworking.

Conformance test is in general not sufficient to actually guarantee that implementations will interoperate successfully. The main reasons lie in the lack of precision of the standards themselves and in the test coverage which can never be exhaustive. Conformance test of communication protocols is directed towards one-to-one relationships, that is the communication between the system under test and a reference implementation in the form of the conformance test system itself. Successful interoperability, however, depends on the communication between a set of implementations in a given application context. Key issues such as test coverage, declaration of test environment and reproducability of test results mean that interoperability testing must be considered as a supplement to and not as a substitute of conformance testing. A relatively complex interoperability test will have little chance of success if the implementations have not been conformance tested in advance.

In CNMA Phase 1 (January 1986 - April 1987) the Fraunhofer Institute, an independent R&D organization in Karlsruhe, Germany, developed or obtained conformance testing technology for CNMA. The protocols tested were the Transport and Session layers as well as Phase 1 Manufacturing Message Specification (MMS). The conformance testing for CNMA Phase 2 (January 1987 - December 1988) was again provided by the Fraunhofer Institute which developed test systems for MMS and NM (Network Management).

The urgent requirement for conformance testing prior to the Enterprise Networking Event ENE '88i held in Baltimore in June 1988 gave CNMA the opportunity to establish a set of unique conformance test tools for MAP/TOP which are now recognized world-wide. A modified consortium was formed for this purpose and the work was carried out in the CNMA Phase 3 from September 1987 to June 1988. The conformance test system development was carried out by The Networking Centre (UK) and the Fraunhofer Institute. The Networking Centre developed test systems for the lower 3 layers of the MAP/TOP profile, whereas the Fraunhofer Institute developed test systems for the layer 7 prtocols MMS, NM and DS (Directory Services). The conformance testing work of CNMA Phase 3 played no small part in the success of ENE '88i and in CNMA itself.

CNMA Phases 1, 2 and 3 made up Project 955 in ESPRIT 1. The CNMA work is being continued within two closely coupled projects in ESPRIT 2:

ESPRIT Project 2617: CNMA,
referred to as CNMA Phase 4, is a sequel to CNMA Phase 2 and has a duration from December 1988 to January 1991.

ESPRIT Project 2292: TT-CNMA
standing for "Testing Technology for CNMA" continues CNMA Phase 3 and runs from November 1988 to July 1991.

More detailed information on CNMA Phase 1 and 2 can be found in (2) and (3), and on CNMA Phase 3 in (1).

2. Conformance Test development in TT-CNMA for CNMA

The scope of TT-CNMA covers upper and lower layer conformance test development, interoperability test development for MMS and Network Management (NM), and a performance measurement study.

Conformance and interoperability testing of the CNMA Phase 4 implementations was conducted at ISW in Stuttgart. As in previous CNMA phases conformance test was mandatory for all vendors. The conformance and interoperability test development was conducted within TT-CNMA according to a Test Development Strategy Document agreed between TT-CNMA and CNMA.

The TT-CNMA developers attended the CNMA technical meetings during the production of the Implementation Guide (IG 4.0), thus ensuring that the test systems met the CNMA requirements. The test system architectures and the test suites were reviewed by CNMA.

The overall sequence of testing at ISW was conformance testing, interoperability testing, interworking testing and application testing. Of these, conformance and interoperability testing were done with test tools provided by TT-CNMA. On the other hand, interworking testing and application testing were pragmatic test phases to meet the pilot requirements. For more information on the interoperability tools for NM and MMS see (5).

Since most applications at ISW depend on MMS for their data communication, MMS conformance test was done first. In addition, testing of the layers Session, Presentation and ACSE was done as part of the MMS test. NM, Directory Service (DS) and MMS Application Interface (MMSI) conformance test followed. Embedded testing within MMS and MMSI conformance test were new to this phase of CNMA. These conformance test tools were all developed in TT-CNMA by the Fraunhofer Institute IITB.

Conformance testing at ISW took place in the period January till June 1990. The conformance test system was provided on two Sun work stations on average; during peak periods three installations were available. The conformance testing was done by the Fraunhofer Institute IITB. In addition, the CNMA vendors were able to use the test systems for preliminary testing and they will receive the conformance test tools for in-house use after the testing at ISW.

The testing progressed well with only minor problems in the test systems, which, with few exceptions, were fixed on site. The test against several independent implementations helped to consolidate and validate the test tools themselves.

Conformance testing uncovered several errors and ambiguities in the specification (CNMA IG 4.0). The corrections were fed into the errata process which led to CNMA IG 4.1. Implementations were subjected to regression testing where necessary to ensure that they complied with the final specification.

TABLE OF CNMA IMPLEMENTATIONS WHICH COMPLETED CONFORMANCE TESTING

Embedded Test of Session, Presentation and ACSE under MMS:

Number of test cases: 92

Implementor	Machine
Bull	Bull DPX 2000
Bull	Bull BM600
GEC	Intel 310
Nixdorf	Nixdorf Targon 35
Olivetti	Olivetti LSX3020
Robotiker	Bull BM600
Siemens	Intel 310

MMS Provider Conformance Test:

Number of test cases: 182

Implementor	Machine
Bull	Bull DPX 2000
Bull	Bull BM600
GEC	Intel 310
Nixdorf	Nixdorf Targon 35
Olivetti	Olivetti LSX3020
Robotiker	Bull BM600
Siemens	Intel 310

MMS Server Conformance Test:

Number of test cases: 80

Implementor	Machine
GEC	GEC PLC
Robotiker	Bull BM600
Siemens	Siemens PLC

MMSI Conformance Test:

Number of test cases: 101

Implementor	Machine
Bull	Bull DPX 2000

NM Agent Conformance Test:

Number of test cases: 268

Implementor	Machine
Bull	Bull DPX 2000
Bull	Bull BM600
GEC	Intel 310
IITB + Bull	Bull DPX 2000
Nixdorf	Nixdorf Targon 35
Olivetti	Olivetti LSX3020
Siemens	Intel 310

NM Manager Conformance Test:

Number of test cases: 102

Implementor	Machine
Bull	Bull DPX 2000
IITB + Bull	Bull DPX 2000
Siemens	Intel 310

DS User Agent Conformance Test:

Number of test cases: 56

Implementor	Machine
GEC	Intel 310
Nixdorf	Nixdorf Targon 35
Olivetti	Olivetti LSX3020
Siemens	Intel 310

DS System Agent Conformance Test:

Number of test cases: 68

Implementor	Machine
Nixdorf	Nixdorf Targon 35
Olivetti	Olivetti LSX3020

3. Conformance test development in SPAG-CCT for OSI-NMF

The OSI Network Management Forum (OSI-NMF) is a large group of IT companies which includes all of the major players in communication networks. The primary goal of the Forum is to make Network Management products and services from different suppliers able to manage communications and complete networks. A number of public demonstrations beginning in the Autumn of 1990 is planned (Forum Show Cases).

The CNMA and OSI-NMF work on Network Management is complementary and will provide major impulses to the integration and standardization of Network Management in enterprises. The two groups have established a technical liaison which has led to the exchange of specifications and alignment where appropriate. This collaboration is expected to be deepened in the upcoming CNMA Phase 5.

OSI-NMF, COS and SPAG (acting on behalf of SPAG-CCT; see below) have reached an agreement on the development and provision of conformance test tools for the OSI-NMF specifications. As a subcontractor to SPAG-CCT, the Fraunhofer Institute IITB has developed conformance testing capabilities for the upper layers of the Forum profile.

This covers systematic testing of the Network Management CMISE/CMIP (Common Management Information Service Elements/Protocol) and embedded testing of ROSE, ACSE, Presentation and Session under CMISE/CMIP with the so-called CMISE/CMIP Tool. In addition, the NM Tool provides test capability for the Forum Objects and Messages.

The CMISE/CMIP and NM Tools were developed using the technology of the TT-CNMA upper layer conformance test systems. Since X.25 is in the Forum profile, an Alcatel-TITN Transport Platform with X.25 connection was integrated into the test tools.

The strategic value and technical scope of the NM conformance testing technology developed originally for CNMA has been greatly enhanced by the cooperation with OSI-NMF.

4. Marketing of conformance test systems by SPAG-CCT

At the end of CNMA Phase 3 in 1988 a consortium of companies interested in testing technology was formed to productize and market the test systems. This consortium called SPAG-CCT (CCT stands for CNMA Conformance Testing) comprises SPAG, ACERLI, Alcatel-TITN, Bull, BMW, Fraunhofer Institute IITB, Nixdorf, ICL and Siemens. The Networking Centre is also involved as a subcontractor. SPAG is marketing agent and manager of the Consortium.

The SPAG-CCT test tools are marketed world-wide. COS and TOYO have been set up as the American and Japanese distributors respectively. COS in the US and JSPMI in Japan have both purchased third party test licences for the

SPAG-CCT tools and are offerring test services.

The test system user interfaces of the SPAG-CCT tools and COS's own tools have been aligned and the tools are available as an integrated tool set.
The current list of SPAG-CCT products comprises conformance test systems for:

the lower layers of the MAP/TOP 3.0 profile

router
802.4 MAC bridge
LLC3
ES-IS
the upper layers of the MAP/TOP 3.0 profile

MMS (Full MAP and Mini-MAP)
NM (Full MAP, Mini-MAP and TOP)
DS (Full MAP, Mini-MAP and TOP)

and as well the CMISE/CMIP and NM Tools for the OSI-NMF Profile.

The lower and upper layer test systems were developed by The Networking Centre and the Fraunhofer Institute IITB respectively.

The upper layer conformance test systems developed by the Fraunhofer Institute consist of a flexible, generic test system for ASN.1 based protocols plus test suites for the specific protocols (4). In this way it has been possible to provide test tools for a range of protocols and profiles. Test tools for further application areas are currently under development.

5. Accreditation and Certification

Conformance testing can gain markedly in long term value if it is supported by a scheme to accredit test laboratories and certify products which is recognized world-wide.

The European Committee for Information Technology & Certification (ECITC) has given official approval to the Recognition Arrangement (RA) of ETCOM (European Testing for Certification for Office & Manufacturing). Test laboratories and certification bodies participating in the ETCOM RA must be accredited to the EN 45000 series documents as applicable. The ETCOM RA will provide requirements specifications which will advise the national accreditation bodies on the technical aspects of the test service operation required of the test laboratory.

The scope of the ETCOM RA currently covers upper layer testing (MMS, NM, DS and embedded testing of Session, Presentation and ACSE) and lower layer testing (router, bridges, ES-IS, MAC and LLC) in the MAP/TOP 3.0 profile.
To date ACERLI in France have obtained accreditation for MMS under the French national scheme and SPAG in Belgium have been accredited for MMS under the

UK NAMAS scheme. In Germany the Fraunhofer Institute IITB has applied for accreditation for MMS with the German DEKITZ which is establishing the necessary procedures in accordance with the ETCOM RA.

Cost is a major hurdle in setting up an accredited test laboratory. The cost arises in part from the equipment, space and test system licences, but can be attributed mainly to the training and retention of the laboratory staff. Indeed, nominated trained staff must be continually available to satisfy accreditation requirements.

As long as there are only a limited number of products, the cost of certification of a given product will be relatively high. Users, when installing a CIM network such as MAP or CNMA, do not usually select (or need to select) a truly multi-vendor system. Moreover, the typical user insists on one vendor or systems house being responsible for the entire system. This focal point of responsibility and obligation cannot, at least up till now and in the foreseeable future, be replaced by a set of product certificates. In fact, product certificates are not expected of vendors in general. Users do not normally have the expertise to test vendor implementations, but naturally still demand high quality products. Vendors will then continue to conduct their own in-house testing in lieu of testing by an accredited laboratory. Large vendors may wish to do their own accredited testing which can lead immediately to a product certificate when required.

6. Outlook

Although the demand for accredited testing is still very low, the demand for advanced conformance test tools, which can be used during the development phase due to their diagnostic capabilities, is high and growing. The conformance test development within and for CNMA has produced a test technology which can be the core technology of test systems in application areas wider than communication networks for CIM.

7. References

(1) BIRTEL, P.: "Conformance Testing for MAP/TOP", presented at Enterprise '88, Baltimore, June 6-9th, 1988.

(2) BOOTH, J. and Girard, B.: "CNMA - Putting Standards to Work", presented at ESPRIT Technical Week, September 28th, 1987.

(3) ROGAN, J.: "Communications for CIM. An overview of ESPRIT Project 955", presented at MICAD '89, Port de Versailles, February 14th, 1989.

(4) WATSON, K.: "Architecture of the Fraunhofer Conformance Test System", pp. 4/167-181 in Proceedings of Enterprise '88, Society of Manufacturing Engineers, Dearborn, June 6-9th, 1988.

(5) WOOG, A.: "Interoperability and conformance in TT-CNMA", ESPRIT Conference Week, November 12-15, 1990.

EMUG

European MAP
Users Group

MAP 3. 0

PRODUCT OVERVIEW

CNMA (OPEN COMMUNICATIONS) CONFERENCE
September 4 - 7, 1990
Stuttgart, Germany

Klaus Grund

E D S
Central European Strategic Business Unit
Rüsselsheim, Germany

ABSTRACT

MAP 3.0 PRODUCT OVERVIEW

MAP - the Manufacturing Automation Protocol - came a long way from the early eighties, when General Motors formed its "MAP Task Force" to find ways of reducing the high cost of manufacturing automation, to MAP Version 3.0 which is based on ISO international standards supporting all seven layers of the OSI Reference Model. Now MAP is recognized as the international standard for industrial communication networks by the leading industrial regions in the world: North-America, Europe (East and West), Japan and Australia.

With the debut of MAP 3.0 at the Enterprise Networking Event in Baltimore in June of 1988, the intend to keep MAP 3.0 stable for 6 years and prices for MAP products reduced, MAP reached a level of maturity such that it can now be used for the implementation of manufacturing automation projects. The most important feature of MAP 3.0 is MMS, the Manufacturing Message Specification, which is a powerful application protocol or language for computer to factory floor device communications. MMS actually makes MAP usable.

Currently, MAP 3.0 is supported by more than 30 vendor companies from the USA, Japan and Europe with products available for most relevant computers and for some important controls covering the area controller, cell controller and device levels of manufacturing applications. The wait for MAP is over now after Digital Equipment, IBM and Siemens finally announced MAP 3.0 support for their products.

Although the set of MAP 3.0 products is still small, it is sufficient to start MAP implementation projects as is evidenced by major installations in North-America (e.g. General Motors, Dupont, Xerox), Japan (e.g. Isuzu Motors, Omron) and in Europe (e.g. Vauxhall, Renault, Volkswagen).

However, before MAP can be used on a larger scale by a wider variety of companies and industries, the following conditions must be met:

1. More PLC, NC machine tool, robot and other automation equipment companies must follow the lead of computer companies and integrate MAP 3.0 and the appropriate functionality of MMS into their products,

2. All MAP products must be conformance tested and tested for interoperability institutions tested for interoperability with key MAP with other MAP products to minimize systems integration costs,

3. Finally, users must include MAP in their strategic networking plans, must make MAP and MMS the basis for their factory automation projects, and must demand MAP products from their favored equipment suppliers.

TABLE OF CONTENTS

1. INTRODUCTION

MAP - the Manufacturing Automation Protocol - came a long way from the early eighties, when General Motors formed its "MAP Task Force" to find ways of reducing the high cost of manufacturing automation, to MAP Version 3.0 which is based on ISO international standards supporting all seven layers of the OSI Reference Model. Now MAP is recognized as the international standard for industrial communication networks by the leading industrial regions in the world: North-America, Europe (East and West), Japan and Australia.

With the debut of MAP 3.0 at the Enterprise Networking Event in Baltimore in June of 1988, the intend to keep MAP 3.0 stable for 6 years and prices for MAP products reduced, MAP reached a level of maturity such that it can now be used for the implementation of manufacturing automation projects. The most important feature of MAP 3.0 is MMS, the Manufacturing Message Specification, which is a powerful application protocol or language for computer to factory floor device communications. MMS actually makes MAP usable.

Currently, MAP 3.0 is supported by more than 30 vendor companies from the USA, Japan and Europe with products available for most relevant computers and for some important controls covering the area controller, cell controller and device levels of manufacturing applications. The wait for MAP is over now after Digital Equipment, IBM and Siemens finally announced MAP 3.0 support for their products. This should make it easier for those control vendors, which do not support MAP yet, to invest in OSI networking.

Although the set of MAP 3.0 products is still small, it is sufficient to start MAP based automation projects as is evidenced by major installations in North-America (e.g. General Motors, Dupont, Xerox), Japan (e.g. Isuzu Motors, Omron) and in Europe (e.g. Vauxhall, Renault, Volkswagen). At the GM Oshawa Car Assembly Plant in Ontario, Canada, MAP 3.0 is already up and running in a large production environment saving money by reducing down times of the factory.

The major driving force for the development of MAP 3.0 products came from the General Motors Saturn Car Assembly Plant in Tennessee, which is a large, brand new facility which will produce GM´s new Saturn cars starting in fall of this year. Saturn uses MAP 3.0 on extensive broadband networks with multiple manufacturing cells and lines on carrierband subnetworks.

The current MAP 3.0 installations are supported by the following leading MAP suppliers: Concord Communications, AEG Computrol, Motorola, Sisco, Hewlett-Packard, Digital Equipment, Allen-Bradley and GE Fanuc.

The MAP 3.0 Product Overview presentation is based on information released by vendor companies at various MAP users group meetings and trade shows in North-America, Japan and Europe. The products mentioned here are either available now, were demonstrated at exhibitions, or will be available in the near future. Not all products listed here are available yet in Europe.

The MAP 3.0 Product Overview presentation is a subset of a report with the same title which is available through EMUG, the European MAP Users Group.

For a complete copy of the latest version of the EMUG MAP 3.0 Product Overview report contact either

Klaus Grund   or   Dave Williams
E D S   EMUG Secretariat
Central European SBU   College of Manufacturing - Bldg. 70
Eisenstrasse 56   Cranfield Institute of Technology
6090 Rüsselsheim   Cranfield, Bedford MK43 0AL
Germany   England
Tel.: +49 - 6142 - 80 23 47   Tel.: +44 - 234 - 75 27 94
Fax: +49 - 6142 - 80 23 47   Fax: +44 - 234 - 75 08 82

2. MAP PRODUCT DEFINITIONS

MAP products can be divided into the following categories:

MAP Boardlevel Products
are boards with 7-layer MAP OSI stack implementation fitting into the backplane of computers or controls based on industry standard buses. Most MAP boardlevel products consist of 2 boards, 1 controller board and 1 modem board supporting broadband or carrierband MAP networks. Mini-MAP implements only OSI layers 1, 2, and 7. MAP/EPA (Enhanced Performance Architecture) nodes support both 7-layer and 3-layer MAP OSI stacks linking Mini-MAP and Full MAP LANs.

MAP Software Products
consist of layer-7 application protocol software, e.g. MMS and FTAM. MMS, the base for MAP 3.0, is available in object code for most standard operating systems and is available in source code for the integration into other computers or

Computers with Integrated MAP Interface
i.e. MAP boardlevel interface and MMS and FTAM implementations.

Controls with Integrated MAP Interface
i.e. MAP boardlevel interface and MMS implementation.

MAP Networking Products
are products which support MAP network infrastructures:
- Headend Remodulators for MAP Broadband networks.
- Bridges to connect MAP Carrierband subnetworks to MAP Broadband.
- Routers to connect MAP networks to other OSI networks like TOP.
- Terminal Servers to interface dumb terminals to a MAP network.
- Network Analyzers / Monitors to analyse / monitor MAP networks
- Network Management Systems to manage large MAP networks.

MAP Migration Products
are box level products which interface existing non-MAP devices and systems:
- Gateways to connect existing proprietary networks to a MAP network, e.g. Allen-Bradley Data Highway, Gould Modbus, Siemens Sinec H1.
- Programmable Network Interface Unit (PNIU), typically based on a PC with a MAP interface including MMS which is programmed to connect any device with a RS-232/V.24 port and proprietary protocol to a MAP network.

MAP Application Products & Tools
are application software packages and tools based on MAP/MMS, e.g. control systems, monitoring systems, data collection systems, development tools, etc.

See Figures A1 & A2 for more information on MAP interface techniques.

3. MAP VENDOR / PRODUCT OVERVIEW

World wide there are more than 30 vendor companies supporting MAP 3.0 by either having products announced, demonstrated or available. The following companies are currently considered to be the leaders in the MAP market:

MAP Boardlevel Products

- Concord   - PC-Bus, PS/2-Bus
- Motorola   - VME-Bus
- AEG Computrol   - Q-Bus, PC-Bus, Multibus, VME-Bus

MAP Software Products

- Sisco   - MMS
- Retix   - FTAM, MMS

Computers with Integrated MAP Interface

- Hewlett-Packard   - HP 9000/800
- Motorola   - Delta Micro Computer
- Digital Equipment   - Micro VAX
- AEG Computrol   - DEC Micro VAX, NCR Tower (Third Party)
- IBM / Concord   - PC/AT, PS/2

Controls with Integrated MAP Interface

- GE Fanuc   - PLC Series 6, CNC Series 15
- GM Fanuc   - R-H Robot
- Allen-Bradley   - PLC-3

MAP Networking Products

- Concord   - Headend, Bridges, Terminal Servers, Monitor
- AEG Computrol   - Headend, Bridges, Monitor
- Motorola / ONE   - Headend, Routers
- Hewlett-Packard   - Network Analyzer
- ITI   - Network Manager

Currently, most of the leading MAP vendor companies are still based in the USA. But there are already initial groups of major European and Japanese companies developing MAP 3.0 products, e.g. AEG, Bull, GEC, Grossenbacher, SattControl, Siemens and Telemecanique in Europe, and Fanuc, Fuji Electric, Fujitsu, Mitsubishi, Omron, Terasaki, Toyo Engineering and Yokogawa in Japan.

See also Figure A9 for an extended MAP 3.0 Vendor / Product Overview.

4. MAP PRODUCTS

Attached tables, <u>Figures A3 - A8</u>, summarize most of the MAP 3.0 products which are currently available on the world market.

4.1 <u>MAP Boardlevel Products</u>

The three leading boardlevel companies Concord, Motorola and AEG Computrol offer MAP 3.0 products covering most industry standard buses (PC/AT-Bus, VME-Bus, Multibus, Q-Bus, PS/2 Micro Channel). Concord supports PC/AT and PS/2 systems, Motorola supports the VME-Bus, while AEG Computrol covers a broad spectrum by offering Q-Bus, VME-Bus, PC-Bus and Multibus products.

Concord reached a major agreement with IBM supplying boardlevel MAP interfaces for IBM PC/AT and PS/2 computers, and 9370 and System/370 type communication frontend processors. Motorola supplies MAP interfaces for most VME-based computers and controls. AEG Computrol´s most visible products are complete MAP interfaces for Micro VAX computers supporting both broadband and carrierband cabling options and full MMS and FTAM application layer protocols. Furthermore, AEG Computrol is the only supplier for Multibus MAP interfaces and also reached an agreement with NCR supplying VME MAP interfaces for Tower computers. Fanuc developed F-Bus MAP interfaces for GE Fanuc CNC and GM Fanuc robot controls.

For a summary of boardlevel products see <u>Figure A3.</u>

4.2 <u>MAP Software Products</u>

This report covers mainly MMS, the main MAP 3.0 application layer protocol, since FTAM plays only a limited role in typical MAP applications. Sisco is the leading company for MMS products supporting most major computer platforms. MMS software products offered by the three leading MAP boardlevel companies - Concord, Motorola, AEG Computrol - are based on Sisco MMS; these companies also offer FTAM based on a Retix implementation.

For a summary of MAP software products see <u>Figure A4.</u>

4.3 <u>Computers with Integrated MAP Interface</u>

Until recently, only two computer companies, Hewlett-Packard and Motorola, and two boardlevel companies, Concord and AEG Computrol supporting IBM and DEC computers, respectively, offered MAP 3.0 computer products. In addition, Motorola and Concord offered third party MAP 3.0 solutions for SUN and Apollo work stations. In October 1989, both IBM and DEC finally announced "native" MAP 3.0 support for their computer systems.

DEC started shipping VAX DEC/MAP 3.0 in December 1989, but initially limits its MAP 3.0 products to Micro VAX computers under VMS and MAP broadband. DEC customers will have to wait till fall of 1990 for DEC MMS software. Sisco MMS has to be used as an interim solution. AEG Computrol has to be considered for MAP carrierband interfaces for Micro VAX computers.

IBM selected Concord as its MAP 3.0 technology supplier and plans to support MAP 3.0 for a wide range of products from PC/AT and PS/2 systems under OS/2 initially to larger 9370 and S/370 type machines under VM in 1990. MMS software developed by IBM will be available in the middle of 1990. For PC/AT and PS/2 platforms Sisco MMS has to be used as an interim solution. Intel's subsidiary Jupiter Technology Corporation offers a third party MAP 3.0 interface for IBM S/370 MVS mainframes supporting FTAM.

In addition, AEG Modcomp, NCR and Bull either announced and / or demonstrated MAP interfaces for their computer systems.

For a summary of computers with integrated MAP interfaces see Figure A5.

4.4 Controls with Integrated MAP Interface

So far only two control companies, Allen-Bradley and GE Fanuc, offered integrated MAP 3.0 interfaces for PLC controls. Siemens is taking the first step towards MAP 3.0 by offering MAP prototype products for PLC controls by mid 1990 and products by the end of this year. AEG Modicon and SattControl initially support Mini-MAP on their PLC controls. There is still limited MAP 3.0 support for CNC and robot controls. GE Fanuc and Grossenbacher so far demonstrated MAP 3.0 for their CNC machine tool controls. The MAP 3.0 interface developed by Fanuc for GE Fanuc CNC controls is also being used for GM Fanuc robots.

For a summary of controls with integrated MAP interfaces see Figure A6. This table also includes information on MAP-Gateways for proprietary industrial LANs and PNIUs which are used to interface non-MAP devices and controls.

4.5 MAP Networking Products

Concord is the leading company supplying a broad line of products for MAP networking, from headend remodulators to bridges, terminal servers and network monitoring tools. Motorola, in cooperation with Open Network Engineering, offers a Router to connect MAP and TOP networks. ITI, the Industrial Technology Institute, has a prototype MAP/TOP 3.0 network manager based on a SUN work station and IBM announced recently that the IBM Systems Manager includes MAP network management functions.

For a summary of MAP networking products see Figure A7. This table also includes information on MAP-Gateways suitable to connect proprietary networks.

4.6 MAP Migration Products

The following companies provide gateways and network interface units to connect non-MAP devices and systems to a MAP network:

4.6.1 MAP Gateways:

Sisco
- MAP Gateway for Allen-Bradley PLC networks
- MAP Gateway for Gould Modicon PLC networks
- MAP Gateway to connect MAP 2.X and 3.0 networks

Square-D
- MAP Gateway for Square-D PLC networks

Siemens
- MAP Gateway for Sinec H1 /AP PLC networks

April
- MAP Gateways for April PLC networks

GE Fanuc
- MAP Gateway / Cell Controller for GE Fanuc controls

Moore
- MAP Gateway for process control proprietary networks

Toyo Engineering
- MAP Gateway for IBM SNA networks

Softing
- MAP Gateway for Profibus (German Field Bus Standard)

4.6.2 Programmable Network Interface Units (PNIUs):

ComConsult - MAP Interface for Controls

ComConsult developed a programmable MAP interface based on a PC/AT under DOS (TARCUS), which can interface up to 4 controls. The initial product interfaces a Bosch CNC over a serial link to a MAP network using MMS.

Procos - MAP Interface Tool Kit for Controls

Procos developed a tool kit based on PC/AT and PS/2 computers under OS/2 and Sisco MMS to implement MAP interface units for PLC, CNC and other controls.

Reflex Manufacturing Systems - MAP Interface for Controls

Reflex developed a shop-hardened programmable MAP interface (MAP Equalizer) based on a single board processor under MTOS and Retix MMS software to interface a variety of non-MAP devices and controls to a MAP carrierband network.

Softing - MAP Interface for Controls

Softing developed a programmable MAP interface for controls based on PC/AT and MMS-Ease. In addition, Softing also offers and supports the Procos MAP Interface Tool Kit for controls (see Procos).

4.7 MAP Application Products and Tools

The following companies provide application products and application development tools supporting MAP and MMS:

Sisco - MAP Application Programs

In addition to MMS-EASE software products for a wide range of computer platforms, Sisco offers the following MAP application products:

- MMS application program tester (F/MAP-TEST)

- MAP device management application program (F/MAP-DMA) for controlling the up- and downloading of programmable devices

- Sisco Easy Access Management System (SEAMS), a menu-driven operator interface for MAP networks

Industrial Technology Institute (ITI) - MMS Device Emulator

ITI has a prototype of a MMS Device Emulator which can emulate MMS messages and communication sequences of typical factory floor devices like PLC, CNC and robot controls. The MMS Device Emulator runs on a PC with a Concord MAP interface. The MMS Device Emulator could be used during the development phase of cell or area controller applications for initial testing without having to connect to real devices.

Lotus Development - Lotus 1-2-3 Spreadsheet

Lotus Development integrated MAP into its spreadsheet software package Lotus 1-2-3 supporting initially the MS-DOS operating systems. With Lotus 1-2-3 MAP it is possible to perform data acquisition at any node of a MAP network and process and analyze that data with the Lotus 1-2-3 Spreadsheet. Lotus 1-2-3 MAP is being distributed and supported by Sisco, Concord and AEG Computrol.

GE Fanuc - Factory Floor Monitoring & Control System

GE Fanuc offers a factory floor monitoring and control system (CIMPLICITY) based on its shop-hardened CIMSTAR CX (MicroVAX II) computers and MAP. This product is based on VAX DEC/MAP 3.0 and Sisco MMS-EASE.

Burr-Brown - Data Collection System

Burr-Brown offers a data collection system (MAP 3.0 Data Collection Factory Network Server) based on a PC/AT as server using a Concord MAP board and Sisco MMS-EASE to connect to a MAP network. Data collection terminals are interfaced to the server over low cost proprietary links.

Procos - MAP Application Programs

Procos developed a set of factory automation and process control software products (EasyMAP) based on MMS, PS/2, PC/AT or industrial computer platforms, OS/2 (UNIX planned) operating systems, Oracle as data base, and Microsoft Excel spreadsheet. EasyMAP supports data collection, report generation, graphic based monitoring and control, and simulation. In addition, Procos offers a tool kit to develop programmable network interface units for non-MAP devices and controls.

Grossenbacher Elektronik - Cell Controller Development Tool

Grossenbacher developed FACTS (Factory Application for Code Generation, Testing & Simulation), a tool based on graphic work stations to develop cell controller applications.

Andersen Consulting - Cell Management Software

Andersen Consulting announced CELL-PAC, an UNIX based open architecture software package that manages the execution of shop floor plans.

Reflex Manufacturing Systems - Area and Cell Controller Software & Tools

Reflex is developing a set of area and cell controller factory automation software packages and tools (CIMPICS) based on UNIX, C, a Real Time Data Base, MAP/MMS, X-Windows, and Grafcet. In addition, Reflex offers a programmable network interface unit to connect non-MAP devices and controls.

Shipstar / MicroBase Software Systems - CASE Tool

Shipstar and MicroBase offer CIMware, a Computer Aided Software Engineering tool based on a PC/AT, for the development of MAP based control systems.

For a summary of MAP Application Software and Tools see Figure A8.

5. MAP INTERFACE COSTS

Since the Enterprise Networking Event in June of 1988, MAP interface prices dropped to a level which made MAP cost competitive with most proprietary networks in multivendor environments. Recent substantial price cuts by Concord and Sisco started a new round of MAP cost reductions.

MAP Boardlevel Products

MAP boardlevel product prices are in the range of $2,300 - $5,000 for broadband and $1,000 - $3,700 for carrierband.

MAP Software Products

MMS software product prices for computers and work stations range from $600 to $1,000 depending on computer and operating systems supported. Most boardlevel product vendors offer FTAM free of charge.

Computers with Integrated MAP Interface

MAP interface costs for computers and work stations including MAP interface board and MMS software range from $1,600 to $9,000 depending on computer size and MAP cabling technique selected. Carrierband MAP interfaces cost about $1000 less than broadband MAP interfaces.

Controls with Integrated MAP Interface

MAP interface costs for PLC controls range from $3,200 to $5,600.
There is no pricing available yet for MAP 3.0 interfaces for CNC and robot controls which are currently under development.

MAP Networking Products

MAP networking products range in costs from $7,500 for a Headend Remodulator, to about $10,000 for a MAP Broadband/Carrierband Bridge, to about $20,000 for a MAP/TOP Router.

MAP Starter Kits

Some of the leading MAP companies offer "MAP Starter Kits", a 3-node MAP network including MAP interface boards for popular computers including MMS software, for users to get started with MAP at special prices starting at $11,000.

Note: Prices given above are in US Dollars. US MAP products outside the USA are more expensive due to shipping, import and value added taxes.

6. MAP PRODUCT DEVELOPMENT TRENDS

The current MAP boardlevel product developments seem to follow the trends stated below:

- Faster on-board processors (32 bit)

- More MAP OSI layer functions on silicon chips

- Single board MAP products (Carrierband)

- More on-board memory (1 - 2 Mbytes)

- MMS and FTAM on board

- Higher performance

  - Larger number of associations
  - Higher data throughput
  - Higher Transaction Rates
  - Lower Response Times

- Additional Cabling Options (*)

  - 802.4 Fiber Optics
  - 802.5 Fiber Optics
  - 802.3 Baseband (Ethernet)

The cost of MAP interfaces is expected to drop with higher production volumes and the integration of additional MAP OSI layers onto silicon chips which will lead to single board MAP interfaces for carrierband as well as for broadband.

Motorola (Tokenbus Controller chip) and Siemens (Carrierband Modem chip) are currently worldwide the leading suppliers for MAP chip level products.

Concord and Motorola predict that MAP boardlevel products will drop to $1,000 for Broadband and to $600 for Carrierband by the end of 1991.

(*) Even though these cabling options are not yet part of the MAP 3.0 Specification, companies like AEG Computrol, Concord and CD Networks (former Thomas & Betts) offer 802.4 Fiber Optic interfaces and companies like Digital, Bull and Hewlett-Packard announced "MMS over 802.3".

7. SUMMARY

As the overview shows there are enough MAP 3.0 products on the market to seriously start MAP pilot and production installations:

1. MAP interfaces including MMS software are available now for most relevant computer platforms and some important controls,

2. MAP migration products (Gateways, PNIUs) are available also to integrate existing non-MAP controls and systems,

3. The first MAP application software products based on MMS entered the market place, and, above all,

4. MAP solutions are becoming more cost competitive with multivendor proprietary networking solutions when looking at total project lifetime costs.

An initial MAP 3.0 pilot project could be started, for example, by selecting from the following equipment with integrated MAP interfaces, some available so far only through third parties (*):

Area / Cell Controllers	• HP 9000/800
	• DEC MicroVAX
	• Motorola Delta
	• IBM PS/2
	• NCR Tower (*)
Work Stations	• Apollo DN 4500 (*)
	• SUN 3/160 (*)
	• IBM PS/2
	• IBM PC/AT or Compatibles
Controls	• Allen-Bradley PLC-3
	• GE Fanuc PLC Series 6
	• GE Fanuc CNC series 15
	• GM Fanuc R-H Robot

See <u>Figure A10</u> for a starter example and <u>Figure A11</u> for a MAP 3.0 pilot example based on currently available products.

A MAP pilot project could be enhanced by employing MAP application software products like Lotus 1-2-3 MAP, Siscos′ SEAMS menu driven MAP interface, Procos′ EasyMAP control system software, ITI′s MMS Device Emulator, etc.

However, before MAP can be used on a larger scale by a wider variety of companies and industries, the following conditions must be met:

1. More PLC, CNC machine tool, robot and other automation equipment companies must follow the lead of the computer companies and integrate MAP 3.0 and the appropriate functionality of MMS into their products,

2. All MAP products must be conformance tested by accredited test institutes and tested for interoperability with other MAP products by systems integrators to further minimize systems integration costs,

3. MAP Network Management and Directory Services must be supported fully by all MAP products to allow large scale MAP installations with centralized network management, and finally,

4. Users now must take full advantage of MAP by:
    *   Including MAP in their strategic networking plans,
    *   Making MMS the basis for their factory automation projects, and
    *   Demanding MAP products from their favored equipment suppliers.

Outlook beyond MAP 3.0

In addition to MMS companion standards for PLC, CNC and Robot controls, and Process Control equipment, which are currently in the process of standardization, the following extensions to MAP 3.0 are considered as possibilities:

*   Fiber Optic Subnetworks

*   Real-Time Enhancements

*   802.3 (Ethernet) Subnetworks

*   Virtual Terminal Support

*   Remote Data Base Access

*

According to the 6-Year MAP Stability Statement, all future MAP 3.0 Release products though must be downward compatible with earlier MAP 3.0 products.

APPENDIX A

MAP PRODUCT DIAGRAMS AND TABLES

TYPICAL MAP 3.0 INTERFACE

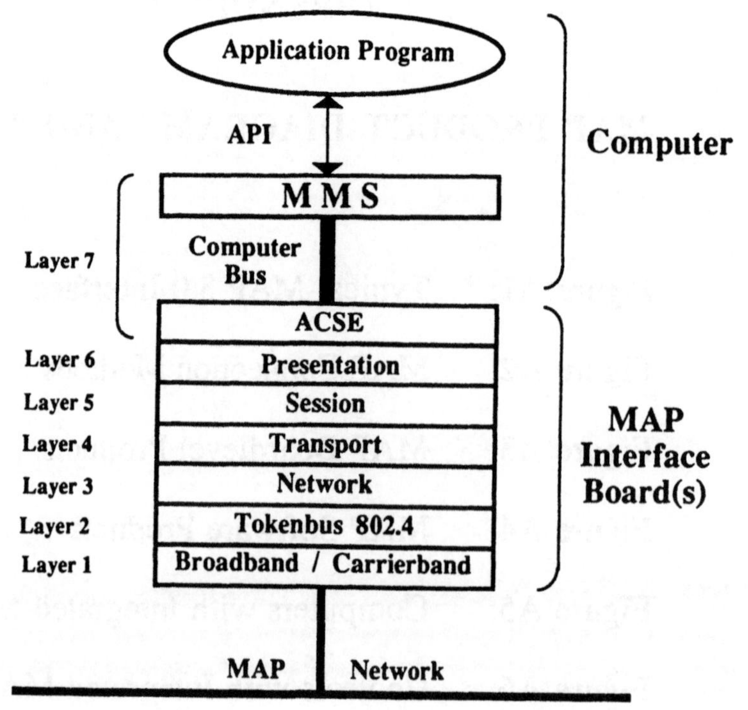

Application Program	
API	Computer
M M S	
Computer Bus	Layer 7
ACSE	
Presentation	Layer 6
Session	Layer 5
Transport	Layer 4
Network	Layer 3
Tokenbus 802.4	Layer 2
Broadband / Carrierband	Layer 1

MAP Interface Board(s)

MAP   Network

MAP CONNECTION METHODS

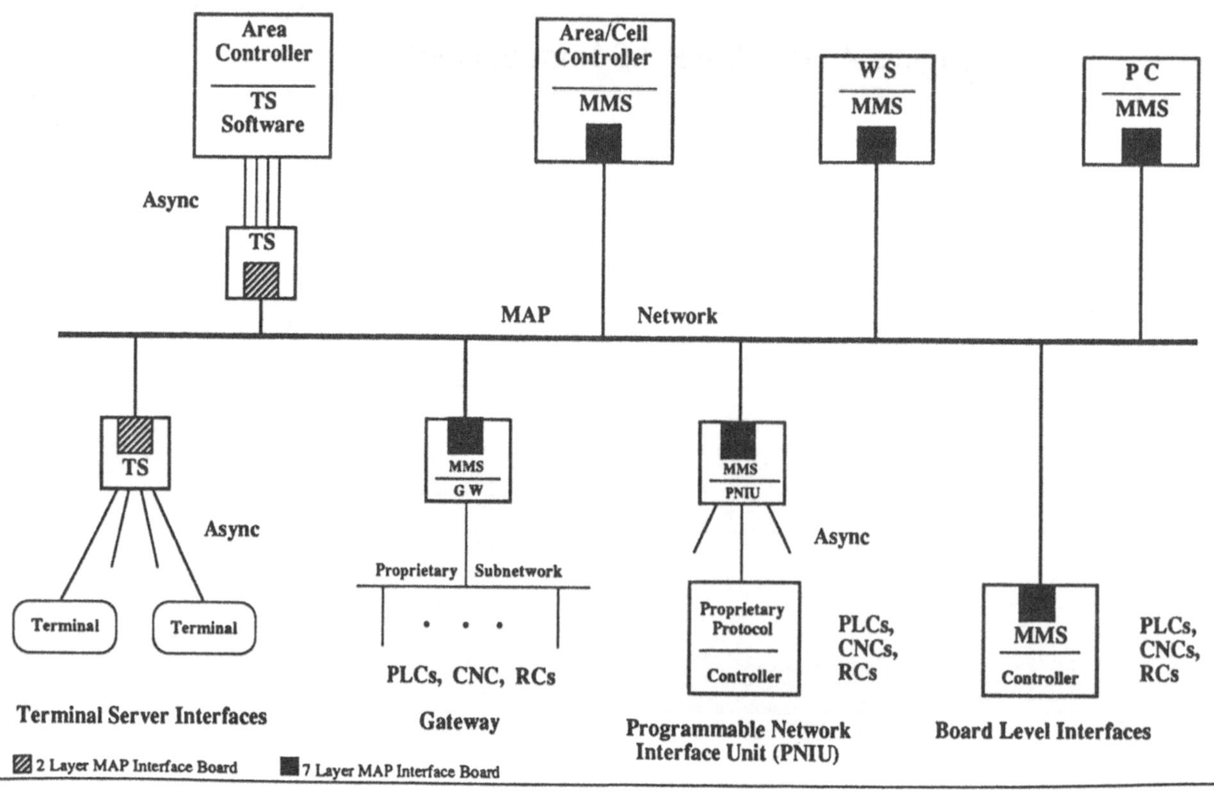

Area Controller — TS Software

Area/Cell Controller — MMS

W S — MMS

P C — MMS

Async

TS

MAP   Network

TS

Async

Terminal   Terminal

Terminal Server Interfaces

MMS / G W

Proprietary   Subnetwork

• • •

PLCs, CNC, RCs

Gateway

MMS / PNIU

Async

Proprietary Protocol / Controller

PLCs, CNCs, RCs

Programmable Network Interface Unit (PNIU)

MMS / Controller

PLCs, CNCs, RCs

Board Level Interfaces

🗗 2 Layer MAP Interface Board   ▪ 7 Layer MAP Interface Board

MAP 3.0 BOARDLEVEL PRODUCTS

(BROADBAND / CARRIERBAND)

Company	IBM		MOT.	INTEL MULTIBUS		DEC	FANUC	NEC	
	PC/AT BUS	PS/2 M.CH	VME BUS	I	II	Q BUS	F BUS	M98 BUS	
AEG Computrol	●		●	●	○	●			
Concord	●	●							
Fanuc							●		
Motorola			●						
Terasaki Electric			●	●				●	

● Available   ◐ Demonstrated   ◉ Announced   ○ Expected in 1991

MAP 3.0 SOFTWARE PRODUCTS (MMS)

Company	IBM MS DOS	IBM PS/2 OS/2	DEC VAX VMS	UNIX	Intel RMX	NEC OS		Comments
AEG Computrol	●		●	●				Based on Sisco MMS Retix FTAM
ComConsult	●		●					Based on Commsoft MMS
Commsoft	●		●					Own MMS
Concord	●	●						Based on Sisco MMS Retix FTAM
Motorola				●				Based on Sisco MMS Retix FTAM
Retix				●[1]				Own MMS Own FTAM
Sisco	●	●	●	●	●			Own MMS
Terasaki Electric						●		Based on Retix MMS

● Available   ◐ Demonstrated   ◉ Announced   1) Source Code Only

243

COMPUTERS WITH MAP 3.0 INTERFACE

Company	Computer	OS	MAP HW	MMS SW	MMS I/F	Status
AEG Modcomp	MC 9730	UNIX	AEG Computrol	Sisco	MMS-Ease	◑
Apollo (*)	DN 4500	AGES/DOS	Concord	Sisco	MMS-Ease	●
Bull	DPX 2000	UNIX	Own	Own		◉
Concurrent Computer	3200	OS/32	Motorola	Sisco	MMS-Ease	◑
Digital Equipment	Micro VAX	VMS	Own	Own	Own	●
GE Fanuc Automation	CIMSTAR DX	VMS	DEC	Sisco	MMS-Ease	●
Fuji Electric	FASMIC G 500	UNIX / OS				●
Fujitsu	A Family & FA PC	SX/A & OS/2				●
Hewlett-Packard	HP 9000/800	UNIX	Own	Own	MMS-I	●
IBM	9370, S/370	VM	Concord	Own	MMS-I	◉
IBM	PC/AT, IC, PS/2	OS/2	Concord	Own	MMS-I	●
Other PC Vendors	PC-XT/AT	DOS	Concord/AEG Computrol	Sisco	MMS-Ease	●
Jupiter Technology (Intel)	S/370	MVS	Intel	---	---	●
Motorola	Delta Series	UNIX	Own	Sisco	MMS-Ease	●
NCR	Tower	UNIX	AEG Computrol	Sisco	MMS-Ease	●
Prime Computers		UNIX	Motorola	Sisco	MMS-Ease	◑
Stratus Computers	XA 2000	VOS				○
SUN (*)	3/160	UNIX	Motorola	Sisco	MMS-Ease	●
Tandem Computers	NonStop CLX	Guardian	Own	Own		◉
Toyo Engineering	TAKE 55	DOS / OS/2	Concord	Retix	Own	●

● Available   ◑ Demonstrated   ◉ Announced   ○ Expected in 1990   (*) Third Party   <u>Note</u>: Most vendors offer also FTAM

CONTROLS WITH MAP 3.0 INTERFACE

Company	PLC	CNC	Robot Control	Cell Controller	MAP Gateway	PNIU	MMS SW
Advanced Computer Comm.		◑					Sisco
AEG Modicon	● 4)				● 3)		Sisco
Allen-Bradley	●				● 3)		Own
April					●		
Cincinnati Milacron		◉	○				
Fanuc		●	●	●	●		
GEC Industrial Controls	◉						
GE Fanuc Automation	● 1)	●		●	●		Own
GM Fanuc Robotics			●				Own
Grossenbacher Elektronik		◑					Own
Mitsubishi	◉	◉		◉			Retix
Omron	◉ 2)			◉			
Satt-Controls	● 2)						
Siemens	◉				●		
Square-D					●		Sisco
Telemecanique	◉						Retix
ComConsult						●	Sisco
Procos						●	Sisco
Reflex Manufacturing Systems						●	Retix
Softing						●	Sisco

● Available   ◑ Demonstrated   ◉ Announced   ○ Expected in 1990   1) Mini-MAP Option   2) Mini-MAP Only
3) Third Party   4) Initially Mini-MAP Only

MAP 3.0 NETWORKING EQUIPMENT

Company	Headend Remodulator	Bridges	Routers	Terminal Servers	Network Analyzer/ Monitor	Network Manager	MAP Gateways
AEG Computrol	●	●			●		
Allen-Bradley	●						
Concord	●	●		●	●		
Fairchild	●						
Fujitsu	●						
Hewlett-Packard		●			●		
IBM						◎	
Industrial Technology Institute						◐	
Motorola / ONE	●		●				
Retix						◎	
Siemens	●				●		●
Terasaki Electric				●			
Toyo Engineering							●
Vance					●		

● Available   ◐ Demonstrated   ◎ Announced

MAP 3.0 APPLICATION PRODUCTS & TOOLS

Company	Application Software	Computer	OS	Status
Andersen Consulting	Cell Controller Software		UNIX	◎
Burr-Brown	Data Collection System	PC/AT	DOS	●
GE Fanuc	Factory Monitoring & Control System	MVAX II	VMS	●
Grossenbacher Elektronik	Cell Controller Software Design Tool	Most WS		◐
Industrial Technology Institute	MMS Device Emulator for Controls	PC	DOS	◐
Lotus (Sisco / Concord / Computrol)	Lotus 1-2-3 Spreadsheet on MAP	PC/AT	DOS	●
Procos	Factory Automation & Process Control Software	PC/AT & PS/2	OS/2	●
Reflex Manufacturing Systems	Area & Cell Controller Software & Tools		UNIX	◎
Sisco	Device Management, Testing, User Interface	PC/AT	DOS	●
Ship Star / MicroBase	Computer Aided Software Engineering Tool	PC/AT	DOS	●
Softing	Factory Automation & Process Control Software	PC/AT & PS/2	OS/2	● [1]
Yokogawa	Process Control System	PC/AT	DOS	●

● Available   ◐ Demonstrated   ◎ Announced   1) Procos Based

MAP 3.0 VENDOR / PRODUCT OVERVIEW

Company	Country	Interface Boards	MMS Software	Computers	Controls	Networking Equipment / Gateways	Application Products
Advanced Computer Comm.	USA				◐		
AEG (Computrol, Modcomp, Modicon)	Germany/USA	●	●	◐	●	●	●
Allen-Bradley	USA				●	●	
Anderson Consulting	USA						◎
Apollo (*)	USA			●			
April	France					●	
Bull	France			◎			
Burr-Brown	USA						●
Cincinnati Milacron	USA				◎		
ComConsult	Germany		●			●	
Commsoft	USA		●				
Concord Communications	USA	●	●			●	●
Concurrent Computer	USA			◐			
Digital Equipment	USA			●			
GEC Industrial Controls	UK				◎		
GE Fanuc Automation	USA/Japan			●	●	●	●
GM Fanuc Robotics	USA/Japan				●		
Grossenbacher Elektronik	Switzerland				◐		◐
Fairchild	USA					●	
Fanuc	Japan	●			●	●	
Fuji Electric	Japan			●			
Fujitsu	Japan			●		●	
Hewlett-Packard	USA			●		●	
IBM	USA			●		◎	
Industrial Technology Institute	USA					◐	◐
Jupiter Technology (Intel)	USA			●			
Lotus Development	USA						●
Mitsubishi	Japan				◎		
Moore Products	USA					●	
Motorola	USA	●	●	●			
NCR	USA			●			
Omron	Japan				◎		
Open Network Engineering	USA					●	
Prime Computers	USA			◐			
Procos	Denmark					●	●
Reflex Manufacturing Systems	UK					●	◎
Retix	USA/Ireland		●			◎	
SattControl	Sweden				●		
Siemens	Germany				◎	●	
Sisco	USA		●			●	●
Ship Star / MicroBase	USA						●
Softing	Germany					●	●
Square-D	USA					●	
Stratus Computers	USA			○			
SUN (*)	USA			●			
Tandem Computers	USA			◎			
Telemecanique	France				◎		
Terasaki	Japan	●	●			●	
Toyo Engineering	Japan			●		●	
Yokogawa	Japan						●

● Available   ◐ Demonstrated   ◎ Announced   ○ Expected in 1990   (*) Third Party

MAP 3.0 STARTER EXAMPLE

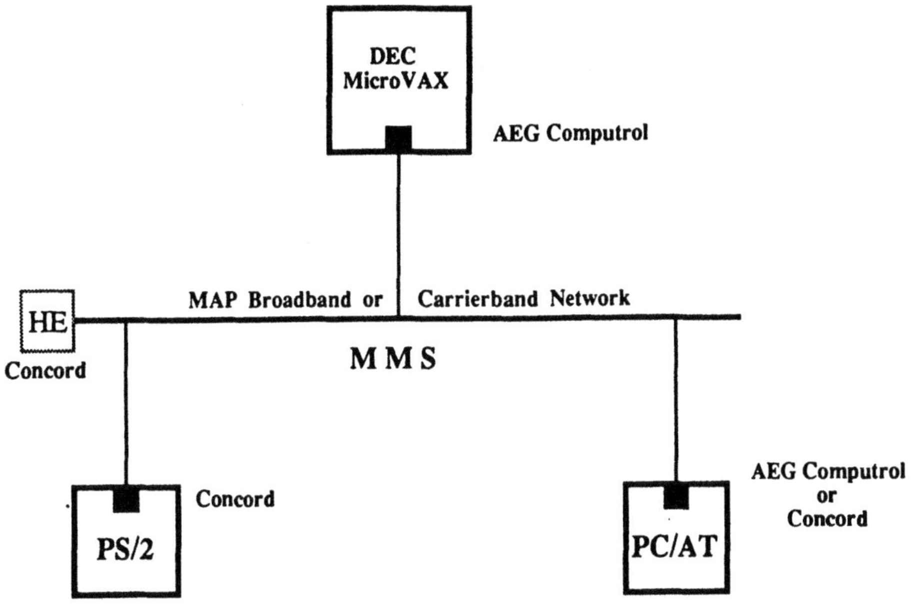

```
                        ┌──────────┐
                        │   DEC    │
                        │ MicroVAX │
                        │     ■    │         AEG Computrol
                        └────┼─────┘
                             │
       MAP  Broadband  or    │   Carrierband  Network
  ┌──┐                       │
  │HE│───────────────────────┴─────────────────────────────┐
  └──┘          M M S                                       │
 Concord         │                                          │
                 │                                          │
                 │                                          │
          ┌──────┼───┐  Concord          ┌─────■─────┐  AEG Computrol
          │   ■      │                    │           │      or
          │          │                    │           │   Concord
          │  PS/2    │                    │  PC/AT    │
          └──────────┘                    └───────────┘
```

MAP 3.0 PILOT EXAMPLE

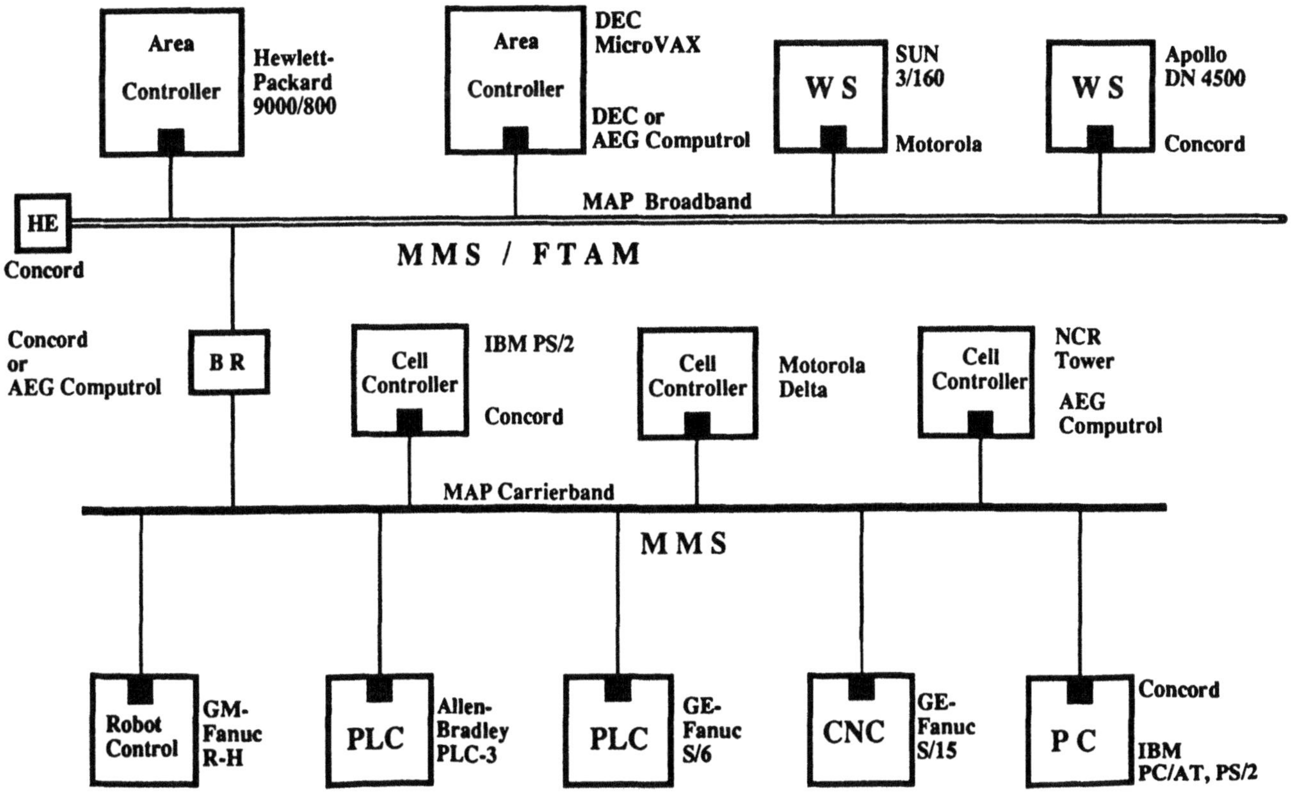

```
  ┌──────────┐              ┌──────────┐              ┌──────────┐            ┌──────────┐
  │   Area   │ Hewlett-     │   Area   │  DEC          │          │  SUN       │          │  Apollo
  │ Controller│ Packard     │ Controller│ MicroVAX     │   W S    │  3/160     │   W S    │  DN 4500
  │     ■    │ 9000/800     │     ■    │              │     ■    │            │     ■    │
  └────┼─────┘              └────┼─────┘  DEC or       └────┼─────┘  Motorola  └────┼─────┘  Concord
       │                         │        AEG Computrol     │                       │
  ┌──┐ │                         │                          │    MAP  Broadband     │
  │HE│─┴─────────────────────────┴──────────────────────────┴───────────────────────┴──────────
  └──┘                          MMS / FTAM
 Concord      │
              │
 Concord   ┌──┐     ┌──────────┐ IBM PS/2   ┌──────────┐           ┌──────────┐ NCR
   or      │BR│     │   Cell   │            │   Cell   │ Motorola   │   Cell   │ Tower
 AEG Computrol└─┼┘     │ Controller│           │ Controller│ Delta     │ Controller│
             │     │     ■    │ Concord    │     ■    │           │     ■    │ AEG
             │     └────┼─────┘            └────┼─────┘           └────┼─────┘ Computrol
             │          │                       │                      │
             │          │        MAP  Carrierband                      │
  ───────────┴──────────┴───────────┬───────────┴──────────┬───────────┴───────────────
                  M M S             │                      │
         │         │                │                      │            │
  ┌──────┼───┐ GM- ┌──────┼───┐ Allen- ┌─────┼────┐ GE-  ┌─────┼───┐ GE-  ┌──────┼───┐ Concord
  │ Robot │  Fanuc│  PLC  │  Bradley│  PLC  │  Fanuc│  CNC  │  Fanuc│  P C  │
  │Control│  R-H  │       │  PLC-3  │       │  S/6  │       │  S/15 │       │  IBM
  └───────┘       └───────┘         └───────┘       └───────┘       └───────┘  PC/AT, PS/2
```

APPENDIX B

MAP 3.0 INSTALLATION SUMMARY

MAP 3.0 INSTALLATIONS - NORTH-AMERICA

- GM Oshawa Car Assembly Plant in Canada
- GM Saturn Car Assembly Plant in Tennessee
- Other General Motors Projects
- Boeing (Planned)
- Eastman Chemicals (Planned)
- Dupont Petrochemical Tank Car Loading System
- Xerox Plant Floor Data Collection System
- Union Camp Paper Mill Process Control System
-

MAP 3.0 INSTALLATIONS - JAPAN

- Isuzu Motors Car Assembly Plant
- Omron Electronic Parts Assembly Plant
- Komatsu Small Press Machine Manufacturing
-

MAP 3.0 INSTALLATIONS - EUROPE

- GM Vauxhall Motors Paint Shop in Ellesmere Port, England

- Volkswagen Car Assembly Plant Paint Shop in Emden, Germany

- Renault Car Assembly Plant in Flins, France

- Aerospatiale Manufacturing System in France

- Esso Automated Oil Drilling Platform Project in Norway (Planned)

- Copenhagen Airport Runway Lightening System in Denmark (Planned)

-

MAP 3.0 PILOT PROJECTS - EUROPE

- British Aerospace CNMA Manufacturing Cell Project in England
- Lucas Diesel Systems DTI Project in England
- National Engineering Lab DTI CIM Project in Scotland
- Manufacturing Cell Projects at Volvo and Saab Scania in Sweden
- Technical University of Stuttgart CNMA Demo Project in Germany
- EDS CIM Center Model Factory in Germany
- Bosch Manufacturing Cell Project in Germany
- KFA Jülich CAD / CAM Project in Germany
- Nuclear Research Center Karlsruhe Robotic Cell Project in Germany
- DISA Project in Denmark
- ITP TUE-TNO Manufacturing Cell Project in Holland
-

EUROPEAN MAP SERVICE COMPANIES

Company	Country	Systems Integration	Inter-operability Testing	Conformance Testing	MMS Portages	Training Programs	EMUG Competence Centers *)
Acerli	France		●	●			
ComConsult	Germany	●	●		●	●	
DataStaff Engineering	France	●			●	●	
EDS	Germany	●	●				
Fraunhofer Institut (IITB)	Germany	●		●			
Institute for Indust. Inform. Tech.	UK					●	
ITP TUE - TNO	Netherland	●	●			●	
KEMA	Netherland			●			
KFA Jülich	Germany	●			●		
PERA	UK	●				●	
Procos	Denmark	●				●	
Reflex Manufacturing Systems	UK	●					
Softing	Germany	●			●	●	
SPAG	Belgium		●	●			
The Networking Centre	UK			●			

*) To Be Established

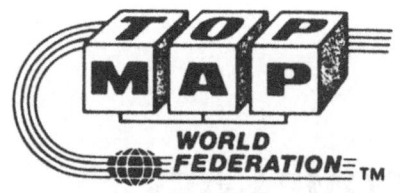

EMUG

European MAP
Users Group

Manufacturing

Automation

Protocol

- INTRODUCTION -

CNMA (OPEN COMMUNICATIONS) CONFERENCE
September 4 - 7, 1990
Stuttgart, Germany

Klaus Grund

E D S
Central European Strategic Business Unit
Rüsselsheim, Germany

INTRODUCTION TO MAP

- MAP / TOP Milestones

- Enterprise Communication Requirements

- MAP / TOP / OSI Reference Model

- MAP / TOP Applications

- MMS Manufacturing Message Specification

- MAP Cabling Options

- MAP Networking Elements

- Enterprise MAP / TOP Networking Model

- MAP Benefits

- MAP for Smaller Companies

- MAP Vendor / Product Summary

- MAP Installation Summary

DEFINITIONS

MAP

- Manufacturing Automation Protocol -

is an OSI Profile for Manufacturing Environments

based on ISO International Standards

TOP

- Technical and Office Protocol -

is an OSI Profile for Engineering and Office Environments

based on ISO International Standards

MAP / TOP MILESTONES

1980 ■ General Motors Formed "MAP Task Force"

■ ISO OSI Seven-Layer Model Specified

1984 ■ MAP Version 1.0

■ First MAP Demonstration at NCC, Las Vegas

■ US MAP Users Group Formed

1985 ■ MAP Version 2.0/2.1 / TOP Version 1.0

■ MAP/TOP Demonstration at AUTOFACT´85

■ US TOP Users Group Formed

■ First Large Scale MAP Broadband Installations

1986 ■ MAP Version 2.2

■ First Large Scale MAP Carrierband Installations

■ First Large Scale MAP Fiber Optics Installation

1986 ■ EMUG - European MAP Users Group Formed

■ OSITOP - European TOP Users Group Formed

■ First European MAP Demo - CIMAP, Birmingham

■ MAP/TOP World Federation Formed

   North-America, Europe, Japan, Australia

1988 ■ MAP Version 3.0 / TOP Version 3.0

■ Enterprise Networking Event ENE´88i, US & UK

■ EMUG MAP Exhibition SYSTEC 88, Munich

1989 ■ Basic Set of MAP 3.0 Products Available

■ First MAP 3.0 Production Installation Running

1990 ■ EEMIG - East-European MAP Interest Group Formed

■ EMUG MAP Exhibition SYSTEC 90, Munich

MAP / TOP WORLD FEDERATION

NAMTUG - North-American MAP / TOP Users Group

EMUG   - European MAP Users Group

JMUG   - Japanese MAP Users Group

AMIG   - Australien MAP Interest Group

EEMIG   - East-European MAP Interest Group

<u>Main MAP / TOP World Federation Goals</u>:

- **Promote MAP / TOP / OSI Standards World Wide**

- **Maintain a Single MAP / TOP Specification**

- **Control Regional Conformance Test & Certification Schemes**

LEADING MAP VENDOR & USER COMPANIES

<u>Over 30 Vendor Companies Support Now MAP 3.0</u>

- **Concord, AEG Computrol, Sisco, Retix, ...**

- **Hewlett-Packard, Motorola, Digital, IBM, NCR, ...**

- **Allen-Bradley, GE Fanuc, GM Fanuc, ...**

<u>Over 20 MAP 3.0 Production Installations (Since 1988)</u>

- **General Motors: Saturn, Oshawa, Vauxhall Motors, ...**

- **Boeing, Eastman Chemicals, Dupont, Xerox, Union Camp, ...**

- **Isuzu Motors, Omron, Komatsu, ...**

- **Renault, Volkswagen, Esso, Copenhagen Airport, ...**

ENTERPRISE COMMUNICATION REQUIREMENTS

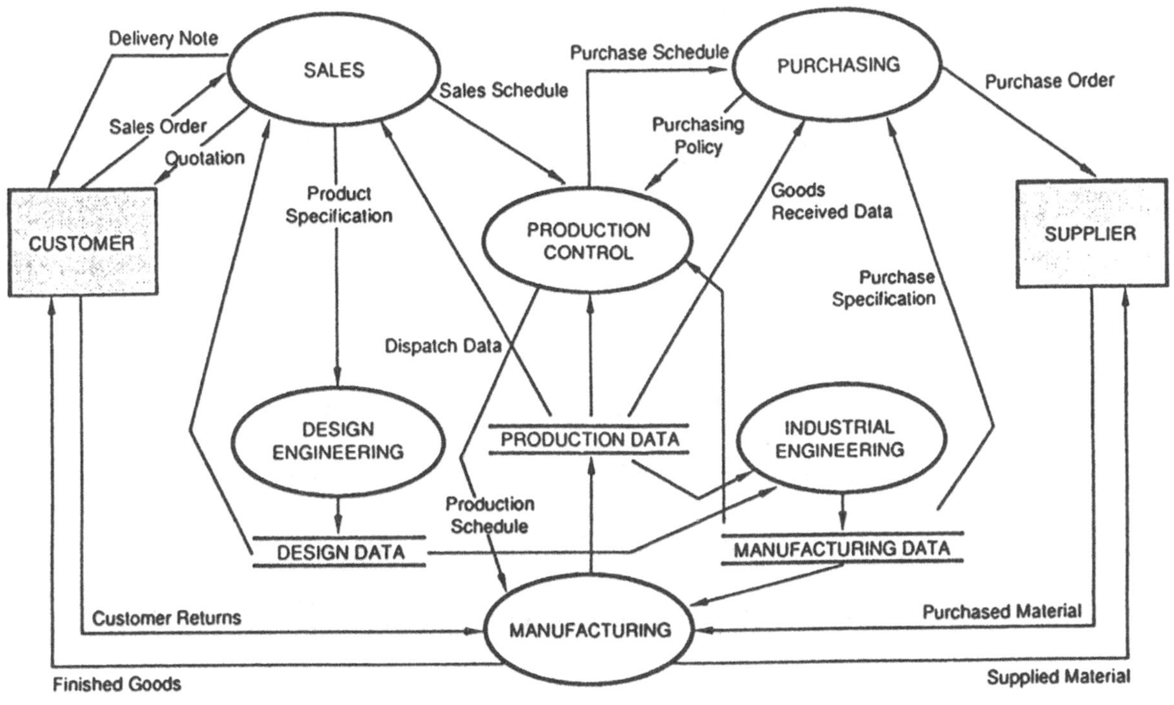

LEVELS OF ENTERPRISE INTEGRATION

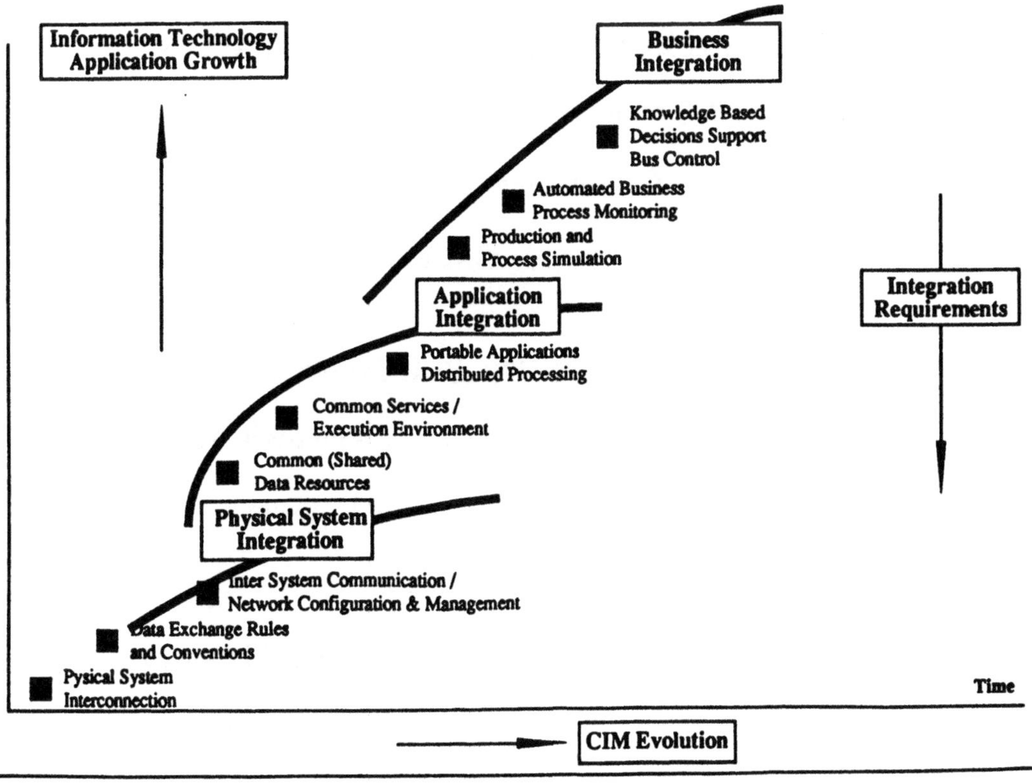

PHYSICAL NETWORK ARCHITECTURES

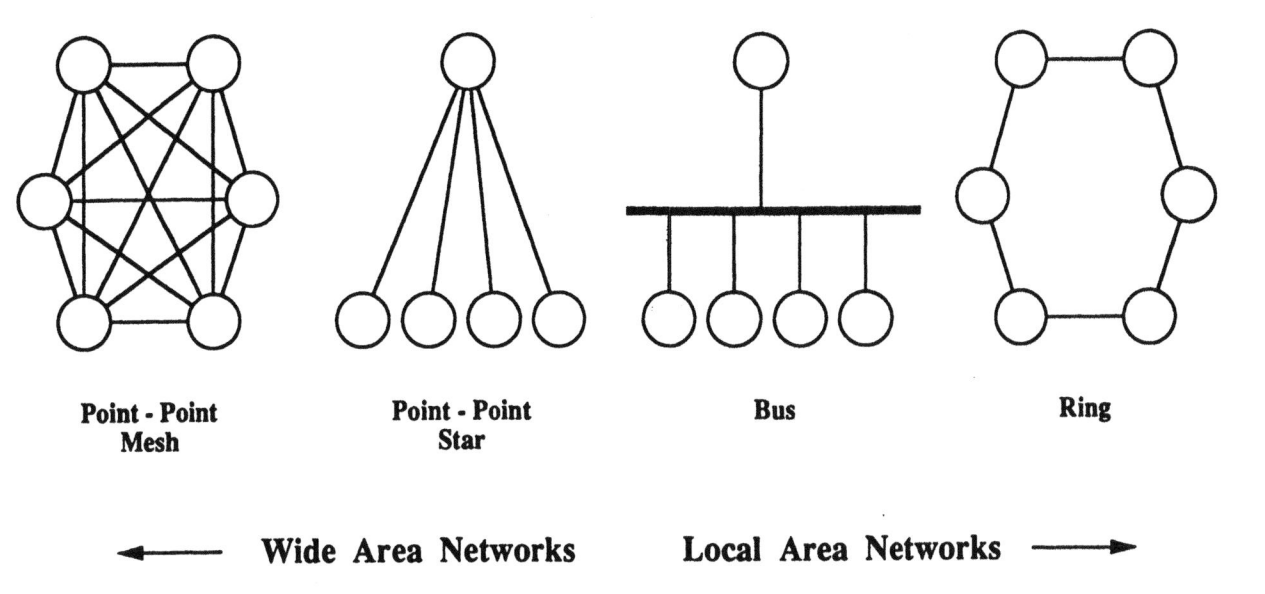

| Point - Point Mesh | Point - Point Star | Bus | Ring |

← Wide Area Networks   Local Area Networks →

WIRING SYSTEMS

- Twisted Pair Copper

- Coaxial Cable

    - Ethernet
    - Cheapernet
    - Carrierband

- Broadband Cable

    - Conventional Cable TV
    - New Cabling System

- Fiber Optics

257

PROPRIETARY NETWORKS

- SNA

- Token Ring

- DECnet

- Ethernet

- Data Highway

- Modbus

- JBus

- Sinec H1

- TCP/IP (Defacto Standard)

-

ISLANDS OF AUTOMATION

Proprietary Networks Generate "Islands of Automation"

Integrating Islands of Automation with Proprietary Networks is:

- Expensive

- Not Flexible

- Not Open

258

ISO / OSI 7-LAYER REFERENCE MODEL

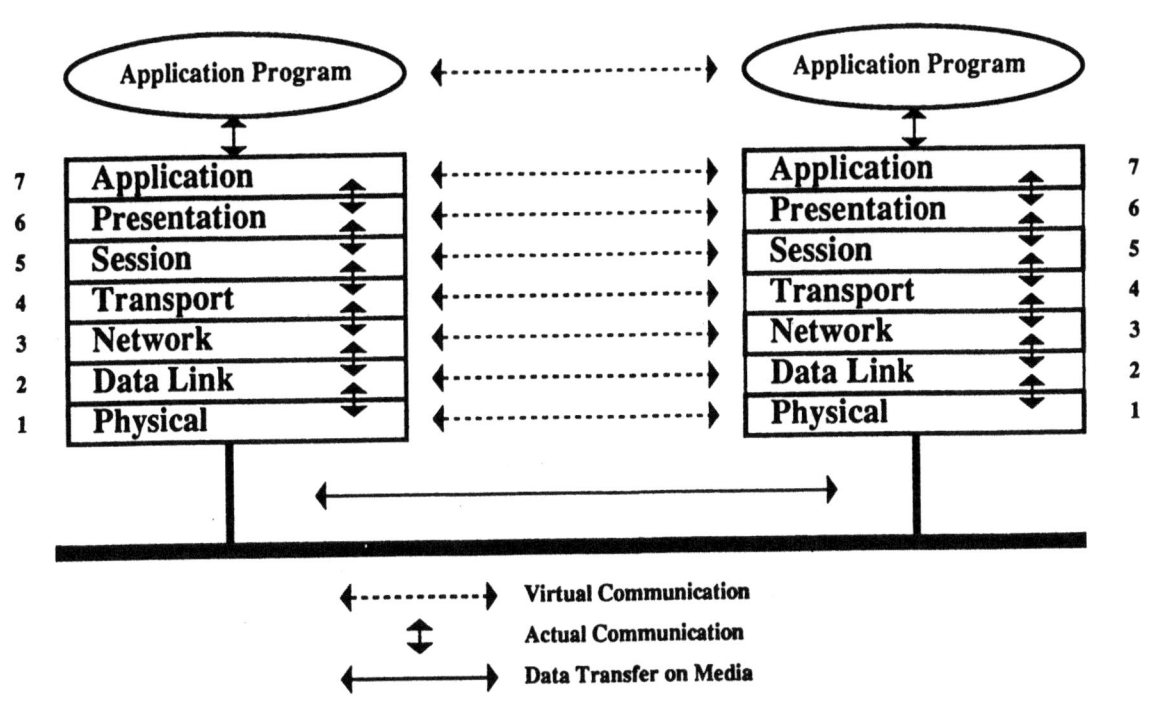

MAJOR MAP GOALS

I. Provide Standard Based Network Solutions for Integrating "Islands of Automation"

II. Provide Standard Based Network Solutions for Connecting Computers with Shop Floor Devices

III. Reduce the Cost of Manufacturing Automation

MAP / TOP 3.0 OSI REFERENCE MODEL

	Factory		Office		Business	
	APPLICATIONS					
	Application		Program		Interfaces	

7	MMS	Network Mgmt	Directory Services	FTAM	Virtual Terminal	MHS
	ACSE					
6	Presentation					X.400
5	Session					
4	Transport					
3	Network					X.25
2	Logical Link Control					
	Tokenbus 802.4		CSMA/CD 802.3		Tokenring 802.5	
1	Carrierband	Broadband		Baseband	Baseband	
	5 MBPS 75 Ohm Coax	10 MBPS 75 Ohm Coax		10 MBPS 50 Ohm Coax	4 MBPS Twisted Pair	

TOP · MAP

MAP / TOP APPLICATION LAYER PROTOCOLS

MMS   -   Manufacturing Message Specification

NM   -   Network Management

DS   -   Directory Services

FTAM   -   File Transfer, Access, and Management

VT   -   Virtual Terminal

MHS   -   Message Handling System (X.400)

ACSE   -   Association Control Service Elements

MAJOR MAP FEATURES

I. <u>Manufacturing Message Specification (MMS)</u>

- ISO Application Protocol Standard for Computers & Controls

- High Functionality (86 Services)

- Standard Application Program Interface (MMS-I)

II. <u>Tokenbus LAN (802.4)</u>

- Broadband for Backbone Networks

- Carrierband for Subnetworks or Smaller Installations

MAJOR TOP FEATURES

I. <u>Multiple Application Protocols</u>

- FTAM

- Virtual Terminal

- MHS (X.400)

II. <u>Local and Wide Area Network Support</u>

- LAN Options: 802.3, 802.5, 802.4 (BB)

- WAN Options: X.25

TYPICAL MAP APPLICATIONS

- Factory Monitoring and Control Systems

- Factory Data Collection Systems

- Distributed Numerical Control Systems (DNC)

- Flexible Manufacturing Systems (FMS)

- Manufacturing Cells and Lines

- Robot Assembly Cells and Lines

- Process Control Systems

TYPICAL TOP APPLICATIONS

Basic Applications

- Remote File Access

- Remote Terminal Access

- Electronic Mail

Data Interchange Format Specifications

- Compound Office Documents

- Computer Graphical Data

- Product Data

MMS - MANUFACTURING MESSAGE SPECIFICATION

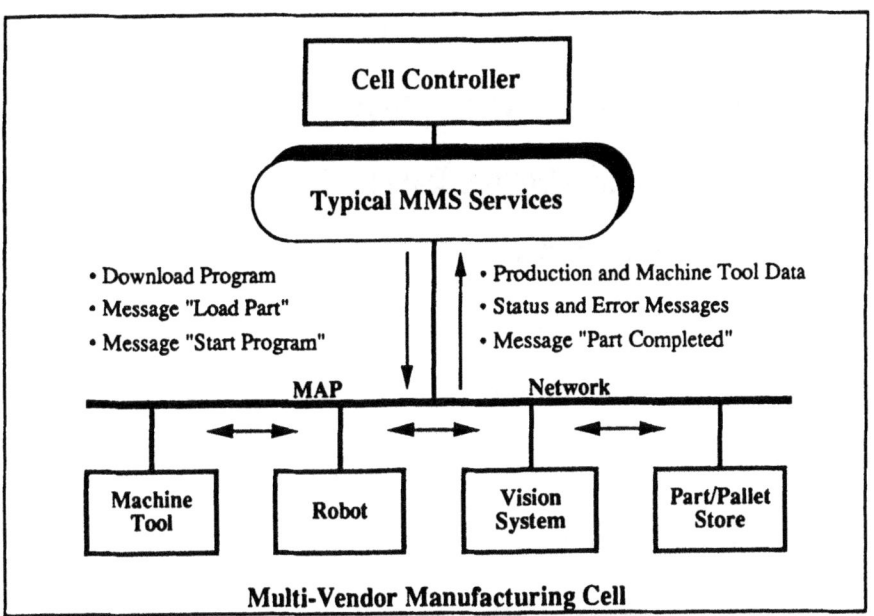

MMS is an ISO Application Protocol Standard for Factory Automation

86 MMS Services with "Conformance Classes" for PLC, CNC and Robot Controls

MMS SERVICES AND THEIR GROUPINGS

COURTESY OF AEG

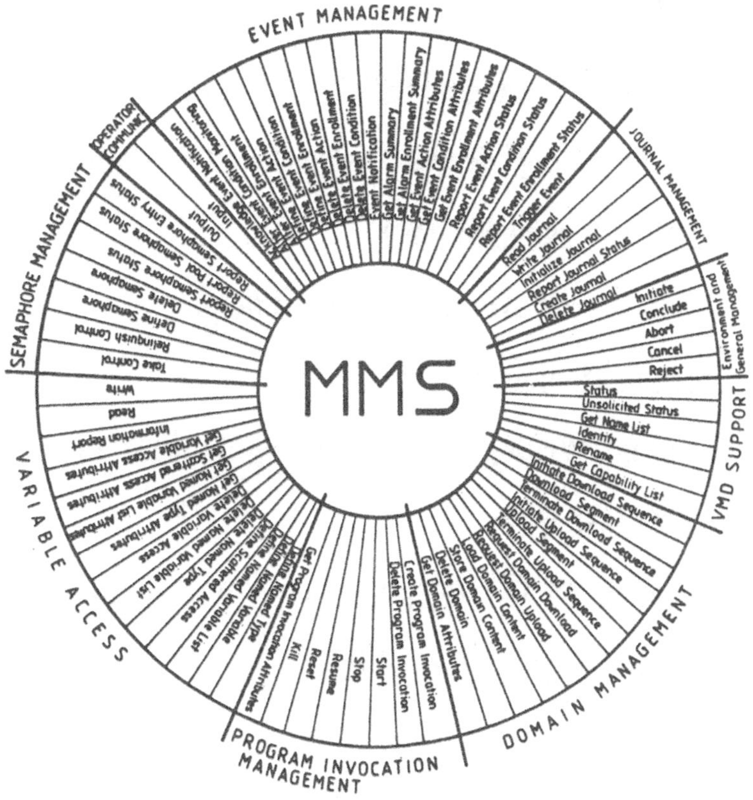

MMS SERVICE GROUPS

- **Environment and General Management** (5)

  Establishing Connections Between Nodes and Applications

- **Virtual Manufacturing Device Support** (6)

  Identification and Status of Virtual Manufacturing Devices

- **Domain Management** (12)

  Up and Downloading of Programs and Data

- **Program Invocation Management** (8)

  Starting, Stopping and Resuming of Programs and Devices

- **Variable Access** (14)

  Defining, Reading and Writing of Variables

MMS SERVICE GROUPS (CONT.)

- **Semaphore Management** (7)

  Sysnchronizing Access Control to Common Recources

- **Event Management** (19)

  Event Driven Automatic Operations

- **Journal Management** (6)

  Reading, Writing and Managing Journal Files

- **File Transfer** (7)

  Subset of FTAM (File Transfer, Access, and Management)

- **Operator Input / Output** (2)

  Simple Console Read and Write Operations to VMD

MMS CONFORMANCE CLASSES

MAP 1 - NC - Machines

MAP 2 - PLC Controls 0

MAP 3 - PLC Controls 1

MAP 4 - Robots

MAP 5 - Process Control 0

MAP 6 - Process Control 1

MAP 7 - Process Control 2

MMS COMPANION STANDARDS

MMS Companion Standards Are Under Development for

- PLC Controls

- NC - Machines

- Robots

- Process Control

MMS Companion Standards Are Device Specific Extensions to Basic MMS

ISO Standard Status for Companion Standards Is Expected by 1990 / 1991

NOTE: Basic MMS Is Usable Without Companion Standards

MAP BROADBAND

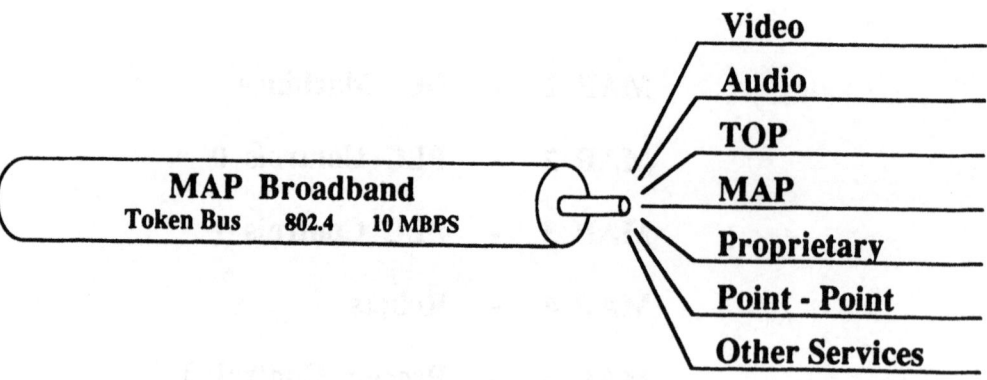

MAP Broadband
Token Bus   802.4   10 MBPS

Video
Audio
TOP
MAP
Proprietary
Point - Point
Other Services

- **Standard 75 Ohm Cable TV Technology**

- **Large Network Capabilities (38 km max)**

- <u>**Choice for Factory Backbone Networks**</u>

MAP CARRIERBAND

MAP Carrierband
Token Bus   802.4   5 MBPS

<u>Single Channel</u>
Data Transmission

- **Same Type of 75 Ohm Coaxial Cable**

- **Maximal 700 Meters, 32 Nodes**

- **No Head End, No Amplifiers, Simpler Modems   -   Lower Cost**

- <u>**Choice for MAP Subnetworks & Smaller MAP Installations**</u>

FUTURE MAP CABLING OPTIONS

- Fiber Optics

    - 802.4 (Tokenbus)

    - 802.5 (Tokenring)

- 802.3 (Ethernet)

802.4 TOKENBUS VERSUS 802.3 CSMA/CD (ETHERNET)

	802.4 Tokenbus	802.3 CSMA/CD
Media Access	Token Passing Deterministic	CSMA/CD Non-Deterministic
Transmission	Modulated	Baseband
Coaxial Cable	75 Ohms	50 Ohms
Max. Lenght	BB: 38 km / CB: 700 m	500 / 2500 m
Channels	BB: Multi / CB: Single	Single
Connections	Tap	Tranceiver
Add Node	Connect to Free Plug	a) Drill Hole - Clamp On or b) Cut Cable - Connect

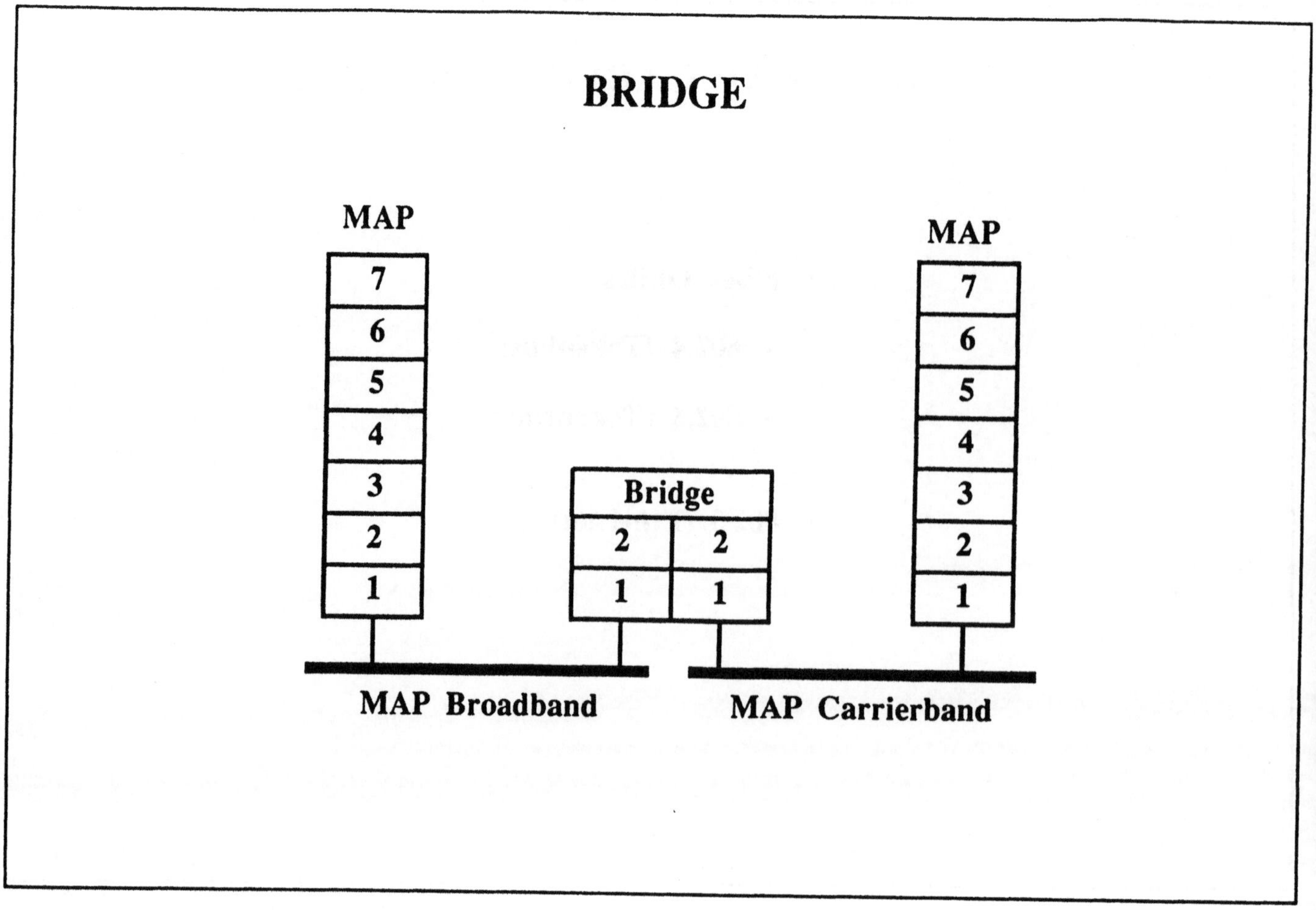

BRIDGE

MAP

7
6
5
4
3
2
1

MAP

7
6
5
4
3
2
1

Bridge	
2	2
1	1

MAP Broadband   **MAP Carrierband**

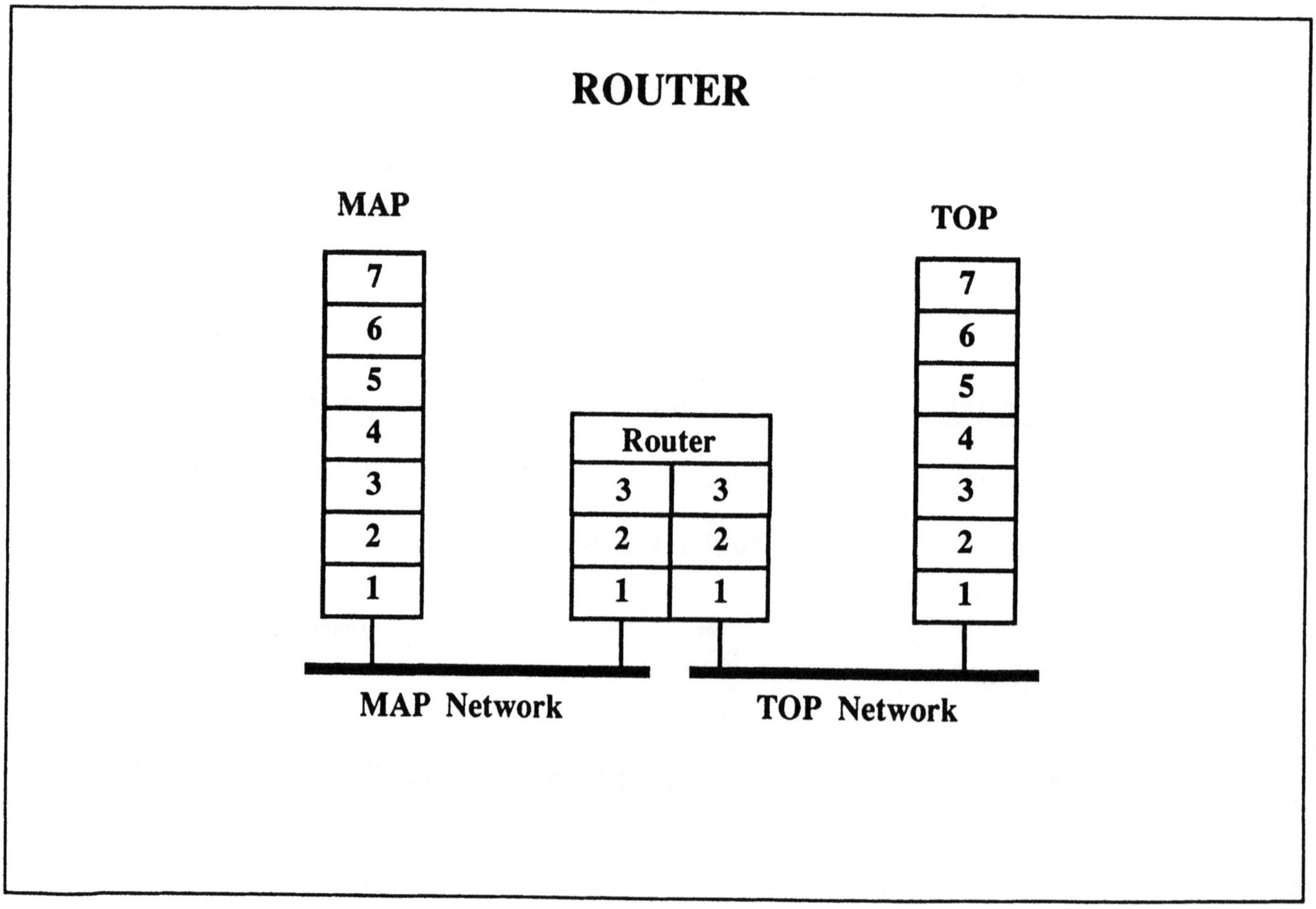

ROUTER

MAP

7
6
5
4
3
2
1

TOP

7
6
5
4
3
2
1

Router	
3	3
2	2
1	1

MAP Network   **TOP Network**

GATEWAY

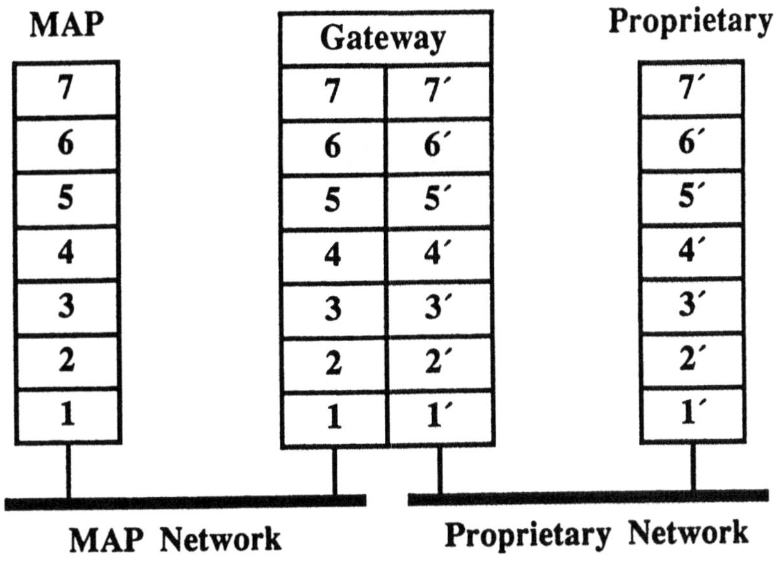

MAP	Gateway		Proprietary
7	7	7′	7′
6	6	6′	6′
5	5	5′	5′
4	4	4′	4′
3	3	3′	3′
2	2	2′	2′
1	1	1′	1′

MAP Network   **Proprietary Network**

MAP / EPA - MINI-MAP

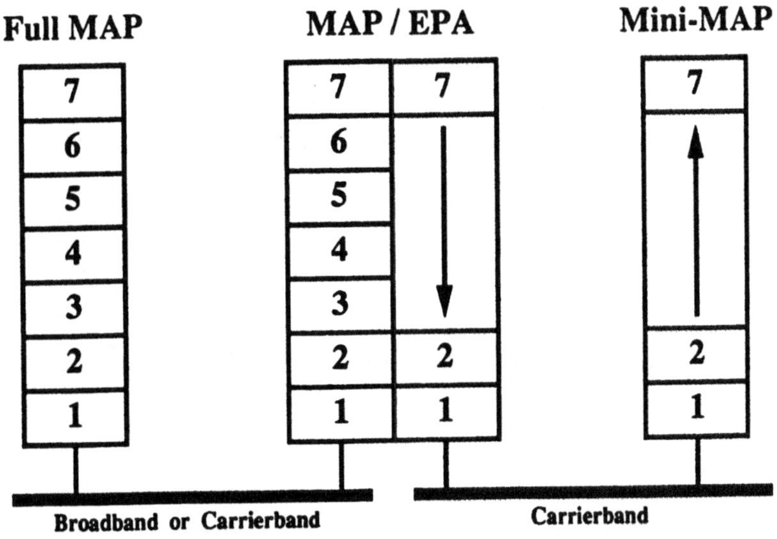

Full MAP	MAP / EPA		Mini-MAP
7	7	7	7
6	6	↓	
5	5		
4	4		↑
3	3		
2	2	2	2
1	1	1	1

Broadband or Carrierband   **Carrierband**

- Full MMS Functionality
- For Backbone Networks

- Full MMS Functionality
- Linking Mini-MAP Subnetworks

- Reduced MMS Functionality
- Faster Response Times
- Lower Cost

EPA - Enhanced Performance Architecture

269

COMMUNICATION FLOWS IN MANUFACTURING

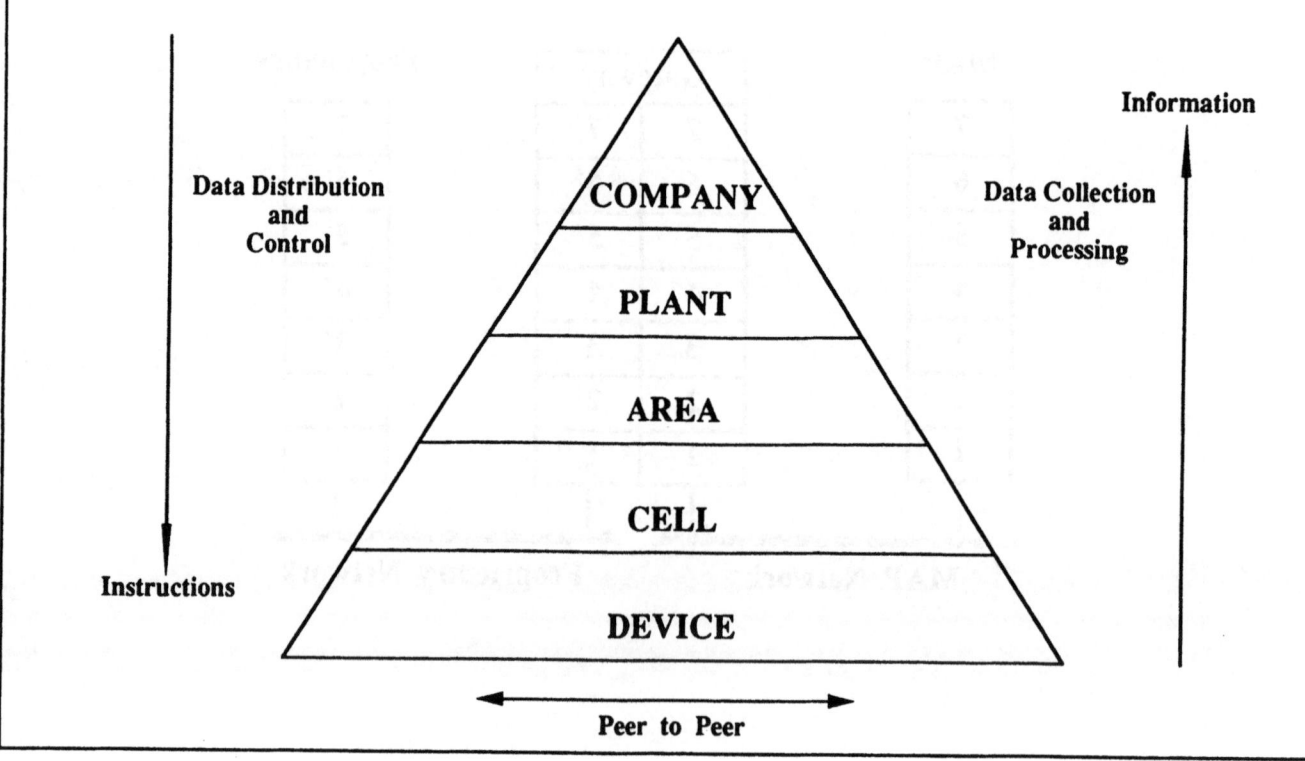

Data Distribution
and
Control

Instructions

Information

Data Collection
and
Processing

COMPANY

PLANT

AREA

CELL

DEVICE

Peer to Peer

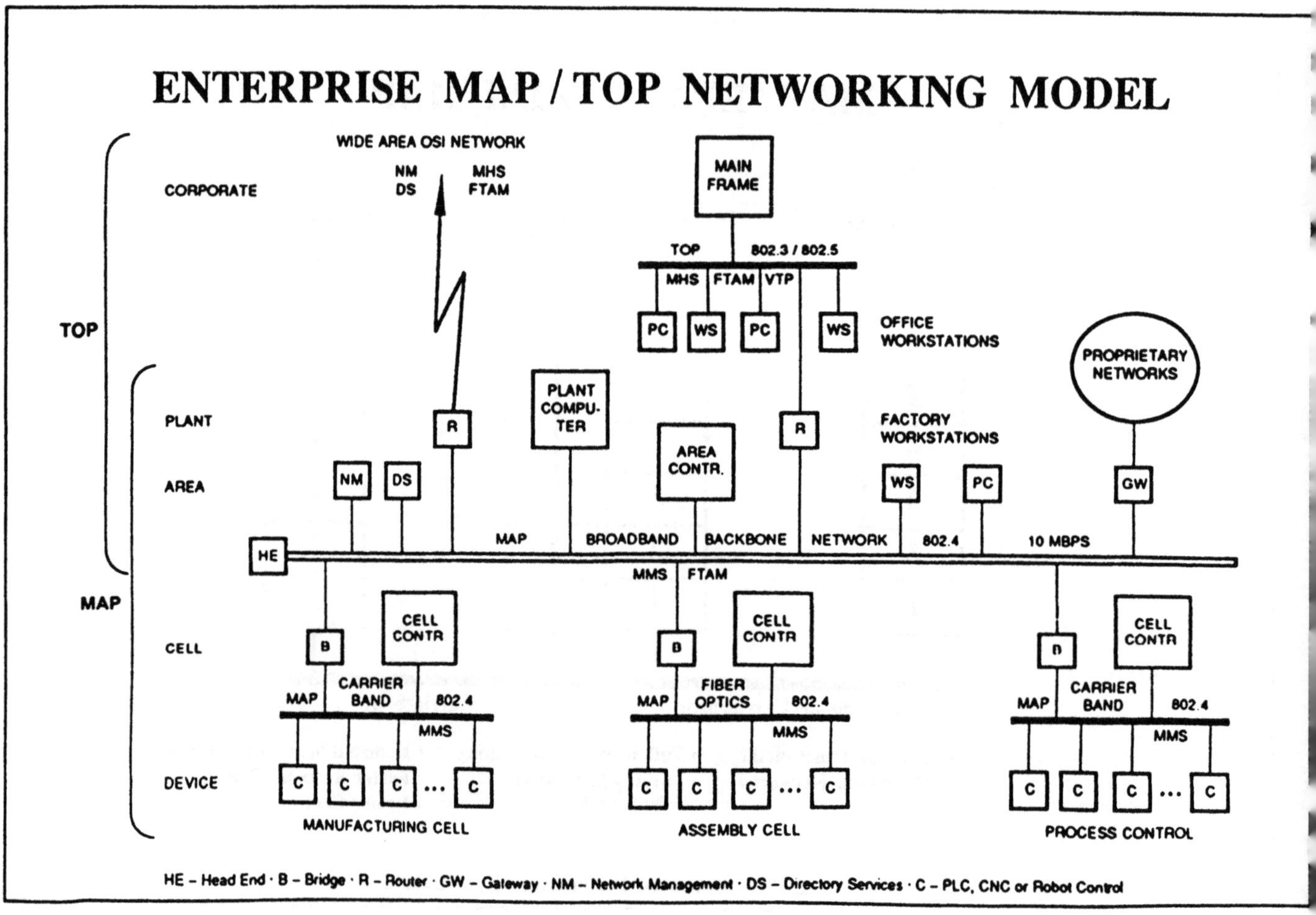

ENTERPRISE MAP / TOP NETWORKING MODEL

CORPORATE

WIDE AREA OSI NETWORK

NM   MHS
DS   FTAM

MAIN
FRAME

TOP   802.3 / 802.5

MHS FTAM VTP

TOP

PC WS PC   WS   OFFICE
WORKSTATIONS

PROPRIETARY
NETWORKS

PLANT

R

PLANT
COMPU-
TER

AREA

NM DS

AREA
CONTR.

R

FACTORY
WORKSTATIONS

WS   PC

GW

HE

MAP   BROADBAND   BACKBONE   NETWORK   802.4   10 MBPS

MAP

MMS FTAM

CELL

B

CELL
CONTR

B

CELL
CONTR

B

CELL
CONTR

MAP   CARRIER
BAND   802.4

MAP   FIBER
OPTICS   802.4

MAP   CARRIER
BAND   802.4

MMS

MMS

MMS

DEVICE

C C C ... C

C C C ... C

C C C ... C

MANUFACTURING CELL

ASSEMBLY CELL

PROCESS CONTROL

HE – Head End · B – Bridge · R – Router · GW – Gateway · NM – Network Management · DS – Directory Services · C – PLC, CNC or Robot Control

MAP BENEFITS FOR USERS

- Users Can Select Best Equipment for a Process

- Vendor Independence

- No More Custom Communications Hardware / Software

- 1 x Training Only (MAP / MMS)

- Lower Systems Integration and Support Costs

- Flexible in Additions and Changes

- Independence from Specialized Support Personnel

RESULT

- <u>Lower Systems Lifetime Costs</u>

MAP BENEFITS FOR VENDORS

- No Duplication of Communication Development Work

- Vendor Can Concentrate on Improving Main Product

- Less Complicated Marketing

- Access to Multi-Vendor Networks and Systems

- Easier to Become a Preferred Equipment Supplier

- Easier Access to World Markets

RESULT

- <u>More Business</u>

MAP FOR SMALLER COMPANIES

VENDOR COMPANIES

- MAP is Especially Important for Smaller Companies

- They Cannot Afford Multiple Communication Schemes

USER COMPANIES

- MAP is Not Just for Large Automotive Companies

- Lower Cost <u>MAP Carrierband</u> is Suitable for Smaller Companies

---

MAP CARRIERBAND NETWORK

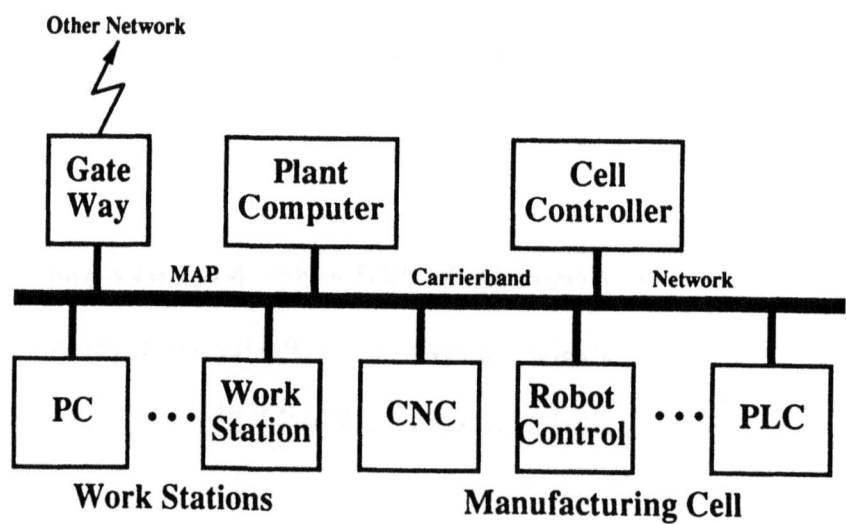

Maximum Length: 700 Meters   No Head End   <u>LOWER COST</u>
Maximum Number of Nodes: 32   No Amplifiers

MAP FEATURE SUMMARY

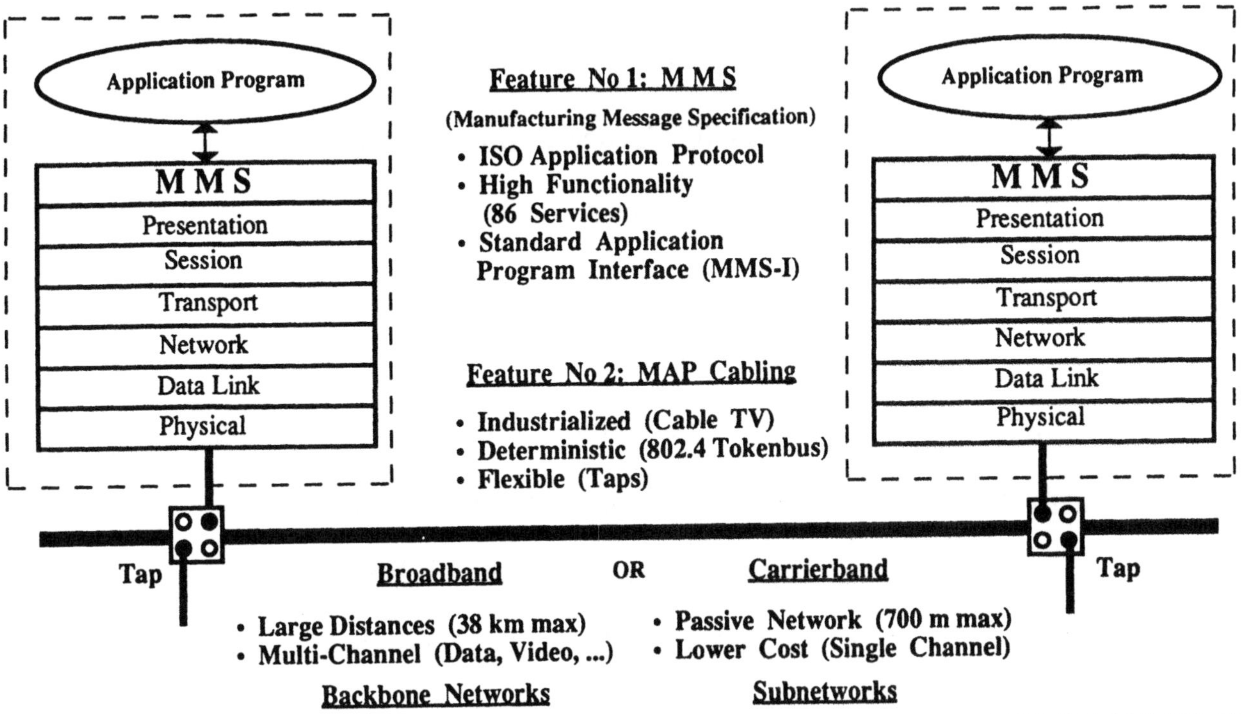

Application Program

M M S
Presentation
Session
Transport
Network
Data Link
Physical

Feature No 1: M M S

(Manufacturing Message Specification)

- ISO Application Protocol
- High Functionality
  (86 Services)
- Standard Application
  Program Interface (MMS-I)

Feature No 2: MAP Cabling

- Industrialized (Cable TV)
- Deterministic (802.4 Tokenbus)
- Flexible (Taps)

Application Program

M M S
Presentation
Session
Transport
Network
Data Link
Physical

Tap

Tap

Broadband   OR   **Carrierband**

- Large Distances (38 km max)
- Multi-Channel (Data, Video, ...)

- Passive Network (700 m max)
- Lower Cost (Single Channel)

Backbone Networks   **Subnetworks**

WHAT IS MAP ?

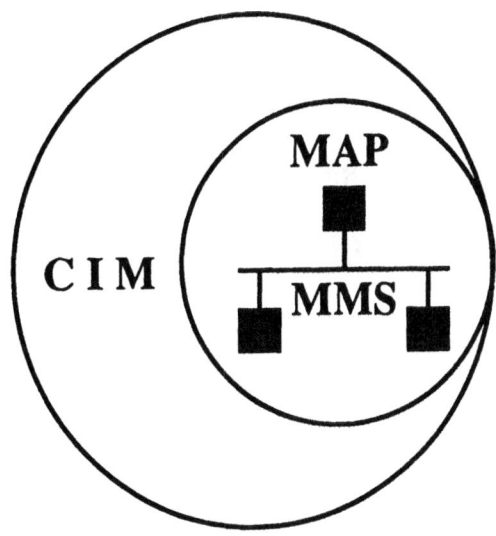

MAP IS A MAJOR BUILDING BLOCK OF **CIM**

MMS IS THE APPLICATION LANGUAGE OF **MAP**

MAJOR MAP 2.X INSTALLATIONS (1984 - 1988)

North-America:
- GM Saginaw Vanguard Factory of the Future
- John Deere & Company
- GMT- 400 Truck & Bus Plants (2)
- Alcoa Aluminum Can Plant
- GM 10 Car Plants (2)
- IBM Endicott

Japan:
- Toyota Automobile Parts Manufacturing Plant

Europe:
- ICL Kidsgrove Plant in England
- Jaguar's Radford Plant in England
- Pilots Projects in England, Sweden, France, Germany, Holland

MAJOR MAP 3.0 INSTALLATIONS (SINCE 1988)

North-America:
- GM Oshawa, Saturn and Other Projects
- Dupont, Xerox, Union Camp
- Boeing, Eastman Chemicals (Planned)

Japan:
- Isuzu Motors, Omron, Komatsu

Europe:
- Vauxhall Motors, Volkswagen, Renault
- Aerospatiale in France
- Esso in Norway, Copenhagen Airport (Planned)
- Pilot Projects in England, Sweden, France, Germany, Denmark, Holland, ...

MAP 3.0 VENDOR / PRODUCT OVERVIEW

Company	Country	Interface Boards	MMS Software	Computers	Controls	Networking Equipment / Gateways	Application Products
Advanced Computer Comm.	USA				◐		
AEG (Computrol, Modcomp, Modicon)	Germany/USA	●	●	◐	●	●	●
Allen-Bradley	USA				●	●	
Anderson Consulting	USA						◎
Apollo (*)	USA			●			
April	France					●	
Bull	France			◎			
Burr-Brown	USA						●
Cincinnati Milacron	USA				◎		
ComConsult	Germany		●			●	
Commsoft	USA		●				
Concord Communications	USA	●	●			●	●
Concurrent Computer	USA			◐			
Digital Equipment	USA			●			
GEC Industrial Controls	UK				◎		
GE Fanuc Automation	USA/Japan			●	●	●	●
GM Fanuc Robotics	USA/Japan				●		
Grossenbacher Elektronik	Switzerland				◐		◐
Fairchild	USA					●	
Fanuc	Japan	●			●		
Fuji Electric	Japan			●			
Fujitsu	Japan			●		●	
Hewlett-Packard	USA			●		●	
IBM	USA			●		◎	
Industrial Technology Institute	USA					◐	◐
Jupiter Technology (Intel)	USA			●			
Lotus Development	USA						●
Mitsubishi	Japan				◎		
Moore Products	USA					●	
Motorola	USA	●	●	●			
NCR	USA			●			
Omron	Japan				◎		
Open Network Engineering	USA					●	
Prime Computers	USA			◐			
Procos	Denmark					●	●
Reflex Manufacturing Systems	UK					●	◎
Retix	USA/Ireland		●			◎	
SattControl	Sweden				●		
Siemens	Germany				◎	●	
Sisco	USA		●			●	●
Ship Star / MicroBase	USA						●
Softing	Germany					●	●
Square-D	USA					●	
Stratus Computers	USA			○			
SUN (*)	USA			●			
Tandem Computers	USA			◎			
Telemecanique	France				◎		
Terasaki	Japan	●	●			●	
Toyo Engineering	Japan			●		●	
Yokogawa	Japan						●

● Available　◐ Demonstrated　◎ Announced　○ Expected in 1990　(*) Third Party

275

EMUG SYSTEC 90 MAP EVENT CALENDAR

22. 10. to   26. 10. 1990   <u>MAP Exhibition SYSTEC 90</u>

"MAP IN EUROPE"

Monday   22. 10. 1990   <u>MAP Tutorial</u>

14:00 - 17:00   MAP For Newcomers

Tuesday   23. 10. 1990   <u>European MAP Forum</u>

09:00 to 18:00

• Keynote Address

• MAP Around The World

• European MAP Installations, Pilots and Strategies

• Highlights of the MAP Exhibition SYSTEC 90

• MAP Vendor Presentations

• Panel Discussions: "The Business Case for MAP"

Wednesday 24. 10. 1990   <u>MAP Workshops</u>

09:00 - 12:00   Workshop 1: MMS and Applications

14:00 - 17:00   Workshop 2: MAP Installation Case Studies

09:00 - 12:00   Workshop 3: MAP Migration Strategies

14:00 - 17:00   Workshop 4: Fieldbus and MAP

RENAULT USER REQUIREMENTS (FOR MAP AND CNMA)

F. Langlois

RENAULT
IGO/ISA service 0484
34 Quai du Point du Jour
BP 103
92109 Boulogne Billancourt Cedex

Abstract

This paper presents the RENAULT company and its associated telecommunications target architecture. Then it focuses on a factory communication and information system.

A case study (example of manufacturing distributed application) is proposed to outline the necessary communication and network management tools, devices, facilities...

A brief synthesis of intended benefits and rationale to use ISO and MAP/CNMA protocols is finally provided.

1. Presentation of RENAULT company objectives

RENAULT is an industrial group, comprising four branches which are :

- automotive (automobile),
- trucks and buses (vehicules industriels),
- finance and services (societes financieres et de services),
- industrial enterprises (entreprises industrielles).

The two first branches represent more than 90% of global turnover.

For all branches, a major objective is a total quality objective.

At the production/manufacturing systems level, this objective is expressed by the following requirements :

- to have a non stop operation (zero fault or a 100% availability) of the information system,
- to adjust performances relatively to the manufacturing process requirements and technology evolutions,
- to control the global system evolution (that means to integrate existing systems with new hardware or software components constituting an improvement without creating any discontinuity, by small but continuous

277

changes),
- to evaluate the quality of service offered by the global data processing system, from a central point, in real time (that means to have the administration tools which are necessary to manage distributed computing systems and the underlying communication networks).

These requirements led to the hereafter CIM recommendations.

2. Enterprise wide communication system

An enterprise network is the result of a telecommunications architecture defined at the CIM "enterprise" level, with extensions to all the enterprise partners (suppliers, public organisations, banks, ...)
RENAULT enterprise network allows the interconnection of the different sites of the enterprise (where each of these comprises a set of interconnected LANs) with WANs facilities.

To classify the sites of the enterprise, three poles of interest are taken into account :

- the administrative pole (headquarters, commercial departments, ...) dealing with administrative activities such as coordination, personnel management, accounting, ...
- the technical pole (research and development departments, methods, CAD/CAM, engineering, ...) dealing with technical activities associated to the conception of the product and the conception of the production tool,
- the industrial pole (manufacturing areas) dealing with production, quality assurance, stocks management, ...

Sites interconnection is based upon a RENAULT private system. All the administrative poles are interconnected via gateways. Most of the technical poles are interconnected. This interconnection is achieved via bridges. The inter-site exchanges relative to industrial poles are either from an administrative nature or a technical nature.

3. Industrial sites communication system

RENAULT sites are categorized accordingly to the three poles previously presented. One site may comprise several poles : a factory comprises all three.

For administrative applications such as administrative messaging, administrative file transfer, transactional access to administrative data bases, virtual terminal, etc. ISO 8802.5 over twisted pair has been selected. SNA is used as a de-facto standard in this application area.

For technical applications, such as graphical (2D, 3D) file transfer, image transfer, scanned documents transfer, file and program transfer, transactionnal interactive access to alphanumerical and graphical data bases, virtual terminal, technical messaging, etc. IEEE 802.3 over twisted pair and/or broadband has been selected. In a short term, ARPA services and protocols are used as an interim solution.

Then in office environment, twisted pair has to be used both for administrative and technical poles' needs.

For manufacturing activities such as industrial messaging, file transfer, virtual terminal, electronic data interchange (for just in time supplying, ...), MAP 3.0 architecture (CNMA 4.0) has been selected.

Within one site, the extended LAN is obtained with the bridged interconnection of all LANs, which allows adaptation of different cabling systems, transmission technics, access methods. The network architecture is a 2-level architecture :
- a backbone LAN is used to federate N dedicated LANs,
- N dedicated LANs according to the standardised LAN
technologies and technics judged the more appropriate.

The systems connected to the network have to provide controlled, reliable and real-time information exchanges between all the information systems of the enterprise and/or partners. Moreover, they have to facilitate information exchange between human beings, so that they may access the required information in the best possible conditions. To perform such a role, the communication system is divided into three independent parts which are :
- the network, described above, comprising an interconnected collection of subnetworks, using intermediate systems,
- value-added services, comprising all the necessary communications protocols to allow information exchanges between distributed application processes (this includes the exchanges of management information),
- the network management application system.

4. Industrial sites information system

The RENAULT CIM architecture, which is described below, is used as a reference for the development of all new manufacturing and/or production applications.

This architecture is based on the following elements :
- as listed before, a two level local communication network conforming to MAP3.0/CNMA4.0 functional profiles,
- data and communication servers shared between operators and applications located in the factory,
- application computers, processing all the necessary information for manufacturing, production monitoring and production control,
- human and mechanical (robots, machine tools, measurement machines, vision systems, AGVs, sensors, actuators, etc...) operators' interfaces used

to allow the communication between :
    . human operators and information systems,
    . mechanical operators and information systems,
    . human and mechanical operators.

Several integration principles have been retained, such as follows :
    - the communication between all interconnected equipments is achieved by using ISO standard protocols such as MMS, FTAM, and other emerging standards such as NFS, SET, SQL, X400, etc...
    - The preferred operating system is currently UNIX system 5, with all its associated graphic interface facilities (X, ...). In some cases where UNIX is not the best solution, or not adapted (for example in case of time critical systems), others operating systems will be selected.
    - Servers and applications are not mixed in the same physical computers, to facilitate maintenance activities and separate evolution.
    - Data servers are intended to maintain all the production and manufacturing data which are going to be shared among all local applications and operators' interfaces. Nowadays, data servers are replicated, but not distributed. In the near future, it is expected that data servers be fully distributed within interconnected computers. For these data servers, relational data bases are used. Currently, INFORMIX has been selected, but other relational data bases such as ORACLE could also be used. Actually, nature and physical location of the data bases will always be hidden to the final users.
    - The existing terminals and computers base will be integrated into this global system by means of communication servers, which are used as gateways between proprietary protocols and ISO ones.
    - Systems and network management tools will be developped in order to facilitate the global system administration from a central point.

The factory information system is a 4-level system comprising :
    - the operator level,
    - the cell level,
    - the shopfloor (or department) level,
    - the factory level.

Information processing, storage and communication resources are more or less distributed within these four levels.

Communication resources are predominating at the operators level, whereas information processing and information storage resources are predominating at the shopfloor/department and factory levels. The cell level comprises the information processing and storage resources allowing a limited autonomy of the cell in case of unavailability of upper level computers.

The two lower levels (operator and cell) are supported by local dedicated communication segments, whereas the two upper levels (shopfloor/department and factory) are supported by the backbone.

5. Case study of distributed manufacturing application

Applications are supported by (micro/mini) computers at the cell and shopfloor/department levels, whereas mainframes are used at the factory level. Production management, manufacturing control and manufacturing monitoring comprise the following applications :
- production planning,
- production scheduling,
- production monitoring, product tracking and traceability management,
- human and mechanical resource management (including personnel management), manufacturing resource planning and maintenance,
- products quality monitoring and management,
- components and modules supplying in some cases on a just in time basis and inventory management,
- manufacturing information and programmes' dispatching on a real time basis,
- manufacturing devices control and monitoring,
- etc...

The degree of criticity of these applications is more or less important relatively to the level of coupling with the manufacturing process, and in the same manner, the real time requirements of the applications are more or less constrained.

The case study hereafter proposed is derived from an industrial application, developped by RENAULT, and currently running in a first factory.
The main goal of this application is to provide the right information to the cars' mounting operators, at the right place and time, under an adapted presentation. In this application, the operators are human operators (about 150 working posts) but in the future, could as well be mechanical operators such as RCs, NCs and/or PLCs, etc..., which in such a case would be down-lineloaded the right programmes and data at the right time.

The human operators can be :
-distributed along the main production line with the task to assemble car components or complete modules on the passing car bodies.
-Locally, in the factory itself, but remote from the main production line, in adjacent working cells, preparing modules which are going to be mounted later on the car bodies, at the main production line level.
-Outside the factory, when considering suppliers working on a just-in-time delivery basis.

The information is sent to all operators, on a real time basis, in a synchronised way with the main production line pace (to give an idea, one car per minute). The schedule of the information transfer is calculated relatively to the nature of the work to be achieved and the speed of the various transportation means (main line conveyors, secondary lines conveyors, AGVs, trucks, etc...) in such a way that the car components/modules arrive at the main production line level in complete synchronization with the car bodies for which they are assigned to.

For example, taking the case of an external supplier working on a just-in-time delivery basis, the information used to confirm delivery orders is sent several hours before the concerned parts be mounted on the right cars, taking into account the necessary delay to load a batch of components/modules, in the right sequence order, on the transportation mean (here a truck), and to transport it at the right place along the main production line.

In order to provide this information in real time, it is necessary to get :
- the production plan of the day, listing the characteristics of the cars which are going to be produced, in the right manufacturing sequence. This information is contained into the production planning machine, which is a mainframe. This production plan may have been transferred in a whole during the previous night, or may be transferred in real time, grouping a batch of vehicles during the production day.
- The bill of materials, components and modules to be supplied for each scheduled car.
- The information of synchronisation which is collected along primary and secondary lines, at checking points level, and is used to identify the cars or components which are passing by, so that it triggers the sending of related information at the right time.

This application contains the following modules :
- the basic modules which are called servers (data servers, communication servers, peripheral/terminal servers) but also cell controllers,
- the application modules.
All these modules are interconnected by means of a standard communication network conforming to MAP3.0.

The data servers modules all together are constituting the products data base of the factory. The current implementation of such a module is based on a UNIX mini-computer using a relational data base (INFORMIX or ORACLE). Several mini-computers are used, in order to guarantee a 100% availability of the system. Currently, data are replicated in different machines, but there is no real distributed system.
This current implementation has been selected for its simplicity, but this doesn't preclude a whole distributed system in the future.
This architecture could also be extended for all kind of data used in a factory such as :
- products data,
- production data,
- production tool data,
- management data,
- etc...
These data servers are not necessarily located in the manufacturing shopfloor, but can also be located in a dedicated computing room.

As well as data servers modules, the communication modules are supported in mini-computers. Their double purpose is :

- to act as a gateway between the existing computer base (IBM mainframe, installed mini-computers non supporting MAP3.0), and the selected standard communication network based on emerging ISO standards (MMS and FTAM). Consequently, they act as integration tools, allowing the integration of the existing factory world within the new architecture based on open standards.

- To act as a gateway between the factory communication network and the public data networks which are used for EDI with the factory's suppliers. Currently X25 plus some proprietary protocols are used. Several studies have been started in order to evaluate new standards such as X400.

These communication servers are not necessarily located in the manufacturing shopfloor, but can also be located in a dedicated computing room.

The peripheral/terminal servers modules are supported by micro or mini-computers, and are located along the production line. This leds to the selection of diskless stations, in order to limitate problems with hardware (problems with disks when used in a harsh environment, with vibrations, dust, etc...). The purpose of these servers is to achieve the interface between data processing equipments (servers and application modules) and human operators who work along the production line. We can find terminal servers when dumb RS232 terminals are used (display, keyboard, badge reader, etc...) but also peripheral servers when more sophisticated stations (workstations) are used (graphic, multi-windowing support, etc...). A terminal server connects dumb terminals with remote computers, in a transparent manner, through an ISO8802.4 standard LAN, and allows to switch these attached devices from one computer to another, either under local operators' request, or under control of the computer it is logically connected to. A peripheral server can be used to locally interface specific devices comprising dedicated protocols.

Cell controllers modules, which are part of the RENAULT CIM architecture, are not belonging to this case study.
These modules are supporting manufacturing devices control and monitoring applications, which are intended to communicate with the case study application described here above, using MMS services. Each of them should control and monitor a group of manufacturing equipments such as :
- PLCs,
- RCs,
- NCs,
- Vision systems,
- Measurement machines,
- AGVs,
- etc...

Application modules are specific purpose processing units which are running one or several dedicated application processes. As an example, the messages dispatching application can be split into the following application processes :

- to collect the information of synchronization provided by sensors, which are distributed along the production line, and to trigger some events when some event conditions are met,
- to schedule the message dispatching relatively to events that have been received, and also to dispatching rules, and to the complete events so that triggering in due time the messages dispatching,
- to prepare the contents of the messages to be sent, relatively to the mounting post activity,
- to dispatch the messages at the right time to the right destination.

Currently, the information of synchronization is provided from a mini-computer through the communication server, which acts as a gateway between the mini-computer proprietary protocol and MMS protocol.

The data which are used to build the messages are fetched by the application modules, in the data servers which are connected on MAP3.0. Planning data are retrieved from the factory mainframe during the night just preceding the considered production day, using the communication server with proprietary mainframe protocol. Then MMS file transfer is used between communication and data servers. The products data, stored within the data servers, can be modified during the production day, if necessary.

6. Network management

From a user's point of view, the network management purpose is to allow him to MASTER its communication network. Actually, this communication network is becoming an important resource of its CIM system upon which all the production of the factories will be rapidly depending. Mastering comprises the following points :
- to have the capability to extend at will the network coverage, the number of connection points, and to increase the traffic relatively to new needs as necessary, keeping the required quality of service level. In summary, to have the capability to take the best profit of new information technologies, and to have the capability of satisfying quickly new communication needs.
- To keep the control of the offered quality of service in terms of network availability and network performance, that is to have a 100% network availability (the production must not stop because of a network fault), and to get the level of performances adapted to manufacturing process requirements, whatever the degree of distributed control.

In order to master a communication network, the network management tools should allow to :
- describe the networks' topology, using previously homologated networks' components,
- identify the relationships between networks' hardware components,
- impose default values to components, in accordance with the site's rules,
- associate above description to real devices,
- configure the networks' components when installation has been successfully completed,
- monitor networks and networks' components in real time, for

their main meaningful characteristics,
- get a global view of the networks' resources status,
- establish, constantly, in real time, the level of quality of service offered, analysing networks' performances,
- achieve some preventive maintenance actions based on trends analysis,
- aid to diagnose when a curative maintenance action is necessary.

7. Intended benefits from ISO and CNMA

The RENAULT telecommunication strategy is based on ISO standards and de facto standards. The main benefits expected from ISO are :
- to facilitate systems integration when deploying CIM concepts,
- to keep a products choice freedom in order to maintain a flexible purchase policy allowing to get the best products/systems satisfying the user requirements at a reasonable cost,
- to lead to a large market allowing mass production and low cost products,
- to reduce the operation and maintenance cost by decreasing the number of technologies and protocols to be managed at a given site,
- to lower the risk of supplying problem by carefully selecting our vendors and possibly having always two sources.

The main benefits expected from CNMA phase 4, in particular, are :
- to impact European communication and network management products via RENAULT user requirements,
- to facilitate relationships with such relative vendors,
- to support RENAULT call for tenders when deploying such solutions in large factories.

Contributor Index